MEDIA DO NOT EXIST

PERFORMATIVITY AND MEDIATING CONJUNCTURES

Theory on Demand #31
Media Do Not Exist: Performativity and Mediating Conjunctures

Authors: Jean-Marc Larrue, Marcello Vitali-Rosati
This book was translated from an original French manuscript by John Detre and Beth Kearney

Cover design: Katja van Stiphout
Design and EPUB development: Valeria Pugliese

Published by the Institute of Network Cultures, Amsterdam, 2019

ISBN: 978-94-92302-39-7
ISBN epub: 978-94-92302-38-0

Contact
Institute of Network Cultures
Phone: +3120 5951865
Email: info@networkcultures.org
Web: http://www.networkcultures.org

This publication is made available through various print on demand services and freely downloadable from http://www.networkcultures.org/publications

institute of
network cultures

CONTENTS

ACKNOWLEDGMENTS

This book is the result of research conducted within the Canada Research Chair on Digital Textualities (http://digitaltextualities.ca), financed by the Social Sciences and Humanities Research Council of Canada and by Le Fonds de recherche du Québec –Société et culture.

Nobody thinks alone: the ideas presented here stem from many years of discussion and collaboration. In many ways, we are not the sole authors. Special thanks go to Peppe Cavallari, Filip Dukanic, Lars Elleström, Erwan Geffroy, Chiel Kattenbelt, Éric Méchoulan, Servanne Monjour, Élisabeth Routhier, Nicolas Sauret, Michael Eberle Sinatra for their assistance, their suggestions, their ideas, their time and their friendship.

Thanks are also in order to other research bodies who supported our writing of this essay: CRILCQ (Interuniversity Research Centre on Québec Literature and Culture), CRIHN (Interuniversity Research Centre on Digital Humanities) and our colleagues from the Centre of Intermedial Research in Arts, Literatures and technologies (CRI/CRIalt) and the International Society for Intermedial Studies (ISIS).

This book has been translated from an original French manuscript by John Detre and Beth Kearney. It was proofread by Beth Kearney. We are very grateful to them for their respectful, attentive and brilliant work which has made these pages possible.

FROM THE BIRTH OF INTERMEDIAL THINKING TO THE CONCEPT OF REMEDIATION: THE MEDIATIC PHASE OF INTERMEDIALITY

Introduction

Henry Jenkins, the keynote speaker at the founding conference of the International Society for Intermedial Studies (ISIS), held in Cluj, Romania in October 2013, opened his remarks by expressing surprise at the turnout. It wasn't false modesty; he was genuinely baffled that the 250 intermedialists in attendance, few of whom were specialists in media or communications, should be interested in his work and that of his colleagues at MIT's Comparative Media Studies Program. The intermedialists hailed from other disciplines in the social sciences and humanities: literature, art history, cinema, music, video, theater, linguistics, history, philosophy, anthropology. Two days later, the author of *Convergence Culture* recovered from his surprise. At the end of the plenary session that closed the first ISIS conference, he remarked, 'I assume that what I have been doing for years is intermediality or has something to do with intermediality'. To the attendees, this was self-evident.

This telling anecdote speaks to a readily observable fact: intermedial research is more common than one might think, but it often appears under other names, and is dictated by disciplinary convention, institutional constraints and editorial policies that occur in the parts of the world where it is conducted.

Born in the wake of what Milad Doueihi calls the *great digital conversion*[1] and what is commonly referred to as the *digital revolution*, intermedial studies are barely 30 years old. That's not much, considering the dispersal and often isolation of the first intermedial researchers, an isolation that was aggravated precisely by the profusion of labels used to describe their subject matter. It took almost 10 years – from the mid-1980s to the mid-1990s – before bridges were built between some of the research groups and the first international research network began to take shape. And it was another 20 years before an international scientific association devoted to intermedial studies was formed with the ISIS founding conference in Cluj, October 24-26, 2013.

While the scattered emergence of intermedial thinking impeded the discipline's growth, it also attests to the supranational and supradisciplinary scope of the intermedial approach, and the ubiquity of the gaps it addresses. Those gaps are not confined to the lightning-fast digital invasion, which has radically transformed our lives since the end of the 20th century, but existed during the pre-digital age, as well. As Espen J. Aarseth observed in one of the foundational texts of game studies,[2] information technology did not make the cybertext possible; cybertexts existed before computers and cybertextual thinking dates back several millennia.

---

1   Milad Doueihi, *Digital Cultures*, Cambridge, Mass.: Harvard University Press, 2011.
2   Espen J. Aarseth, *Cybertext: Perspectives on Ergodic Literature*, Baltimore, Md: Johns Hopkins University Press, 1997.

In support of this assertion, Aarseth shows how the famous Chinese book of changes, the *I Ching*, which dates back to the Zhou dynasty (1027-256 BC), was already a cybertext. The intermedial approach likewise investigates processes that extend into the distant past. Researchers such as Peter Boenisch[3] argue that intermedial processes have existed since the invention of the alphabet, and interartistic theoreticians[4] regard the complex, ceaseless relationships between artistic disciplines since antiquity as compelling models of intermedial dynamics. One of the first propositions advanced by the intermedialists was that the arts are media, indeed the oldest media, and that intermedial phenomena date back tens of millennia.

The term intermediality first gained currency in Germany, Belgium, the Netherlands and Canada. Germany may be considered the birthplace of intermedial thinking, while Canada was the site of the first scientific research center devoted to it, the Centre de recherches intermediales/Centre for Intermedial Research (CRI) at the Université de Montréal, founded in 1997. However, for the reasons we have mentioned, the term 'intermediality' remained confined to a small circle of researchers, even as studies in the field, such as those conducted by Henry Jenkins at MIT, proliferated. In other words, intermedial reflection developed steadily and was constantly enriched, unbeknownst to the researchers themselves. A few examples: in 2000, the American researchers Jay David Bolter and Robert Grusin published what would be considered a groundbreaking work in intermedial studies: *Remediation. Understanding New Media.*[5] The concept of remediation they develop is at the heart of intermedial dynamics. But nowhere in their 295-page tome do the words 'intermedial' or 'intermediality' appear. The same is true of other seminal texts in the evolution of intermedial thinking in the US since the late 1980s. In *When Old Technologies Were New,*[6] which may be considered one of the foundational texts of American intermedial studies, Carolyn Marvin discusses how, as the use of electricity became widespread in the 19th century, electrical media emerged by cannibalizing each other, transforming interpersonal relationships and social behaviour in the process. Her book came out at the same time as the first essays by German intermedialists on related themes, but they do not appear to have been aware of each other.

In 1999, as digital technologies were invading the theatrical stage, Philip Auslander's icon-oclastic book *Liveness: Performance in a Mediatized Culture*[7] took aim at the deep-rooted idea that the direct physical co-presence (in the same place, in the flesh) of senders and

---

3   Peter Boenish, 'Mediation Unfinished: Choreographing Intermediality in Contemporary Dance Performance', in *Intermediality in Theatre and Performance*, ed. by Freda Chapple and Chiel Kattenbelt, Amsterdam; New York: Rodopi, 2006, pp. 151-66.

4   Claus Clüver, 'Intermediality and Interarts Studies', in *Changing Borders: Contemporary Positions in Intermediality*, ed. by Jens Arvidson, Lund: Intermedia Studies Press, 2007, pp. 19-37; Water Moser, 'L'interartialité: pour une archéologie de l'intermédialité', in *Intermedialité et socialité: histoire et géographie d'un concept*, ed. by Marion Froger and Jürgen E and Müller, Münster: Nodus, 2007, pp. 69-92.

5   J. David Bolter and Richard A Grusin, *Remediation: Understanding New Media*, Cambridge, Mass.: MIT Press, 2000. The title echoes the resounding essay published by Marshall McLuhan, *Understanding Media*, New York: Signet, 1966.

6   Carolyn Marvin, *When Old Technologies Were New: Thinking About Electric Communication in the Late Nineteenth Century*, Nachdr., III, 1990.

7   Philip Auslander, *Liveness: Performance in a Mediatized Culture*, London; New York: Routledge, 2011.

receivers in a venue such as the theater and technological mediatization were irreconcilable. Auslander argued that those distinctions born of the operation of electrical media in the late 19th century cannot withstand the reality test: liveness may be mediatized and still be live. In 2003, Jonathan Sterne published another key intermedial work: not only did *The Audible Past*[8] revolutionize sound studies, but it also expanded intermedial analysis to the world of sound, uncovering the full complexity of mediation processes and underlining the hazards of transposing concepts from visual studies to sound studies. In *Always Already New: Media, History, and the Data of Culture*,[9] Lisa Gitelman questions whether what were then referred to, by researchers and in society in general, as 'new media' in the industry really were new. Adapting and developing Derrida's well-known phrase, she contends that all media were always already new. Gitelman's analysis was grounded in a synthetic definition of media to which no intermedialist, in the US or elsewhere, would have objected at the time. She also clarifies and enriches, without naming it, the central concept of remediation, while Alexander R. Galloway questions it in *The Interface Effect*[10] and Craig Dworkin investigates modes of mediation in *No Medium*.[11] In the same year as the publication of Galloway's *The Interface Effect*, Erkki Huhtamo and Jussi Paikka's *What Is Media Archaeology?*[12] marked the arrival of media archaeology as a recognized scientific discipline. Archeology is one of the main branches of intermediality and the remediation model is one of its cornerstones.

Other important texts in narratology and studies of adaptation, reception, electroacoustics, literature and the early cinema published during this period also adopted an intermedial perspective, but none of those books and articles that so shaped the development of thinking about intermediality, not only in the US but around the world, contained the word 'intermediality'.

The same observation can be made about intermedial studies in the French-speaking world, with the exception of research activities in Canada around the Centre de recherche sur l'intermédialité at the Université de Montréal. Works by Philippe Marion on linkages between media,[13] by François Odin on the 'births' of cinema, by Jean-Luc Déotte on the concept of apparatus[14] and by the Toulouse group[15] on the concept of device explore issues in intermediality without calling them by this name. Texts by Jacques Rancière on the medium and,

---

8   Jonathan Sterne, *The Audible Past: Cultural Origins of Sound Reproduction*, Durham [N.C.: Duke university press, 2003.

9   Lisa Gitelman, *Always Already New: Media, History, and the Data of Culture*, The MIT Press, 2008.

10   Alexander R Galloway, *The Interface Effect*, Cambridge, UK; Malden, MA: Polity, 2012.

11   Craig Dworkin, *No Medium*, 1st MIT Press paperback edition, Cambridge, Mass London: MIT Press, 2015.

12   *What Is Media Archeology*., ed. by Erkki Huhtamo and Jussi Parikka, Cambridge: Polity Press, 2015.

13   Philippe Marion, *L'année des médias 1996*, 1997 https://dial.uclouvain.be/pr/boreal/object/boreal:83077 [accessed 16 February 2019].

14   Jean-Louis Déotte, *Appareils et formes de la sensibilité*, L'Harmattan, 2005.

15   Philippe Ortel, 'Note sur une esthétique de la vue: Photographie et littérature', 2002 https://doi.org/http://isidore.science/document/10.3406/roman.2002.1164; Arnaud Rykner, *Le tableau vivant et la scène du corps : vision, pulsion, dispositif*, Hyper Article en Ligne – Sciences de l'Homme et de la Société, 2013 https://isidore.science/document/10670/1.0jcfgt [accessed 16 February 2019].

more recently, on the (emancipated) spectator,[16] by Daniel Sibony on the in-between[17] and by Deleuze on cinema (particularly the crystal image)[18] made such important contributions to intermedial studies that they would certainly warrant inclusion on any list of essential readings in intermediality, but none were, at the time, published under the banner of 'intermediality'.

This first observation leads us to a second. To this day, in the minds of many observers of the social sciences and humanities in general, and the field of arts and media in particular, intermediality boils down to the junction of artistic practices, most often in the form of a technology transfer from one practice to another: using projected images in theatrical productions, turning movies into video games, using photographs in literary texts, and so forth. Certainly, these are all examples of intermedial dynamics, but intermediality is much more than this, as we will show.

Historical Overview

For nearly 20 years, up to the mid 2000s, intermedial research was dominated by a fairly traditional conception of media, which we shall discuss in a later chapter. Intermedial studies sprang from a paradox that digital technology had lain bare: not only do media constantly interact, but they are themselves the products of such interactions. This singular dynamic became the focus of the intermedialists, who sought to understand how, by what means and to what end it was produced. But over the past 10 years or so, the picture has grown murkier as the very concept of media has been challenged; indeed, whether such a thing exists or has ever existed is questionable. The relatively brief history of intermediality can therefore be divided into two periods: the mediatic period and the postmediatic period. The first was founded on a relatively traditional conception of media inherited from Marshall McLuhan and communication theory. The second wrestled itself free from that frame. The transition from one to the other illustrates intermediality's arduous but successful struggle for emancipation from the poststructuralist mindset from which it emerged.

As we understand it, intermediality refers at once to a field of study, a dynamic and an approach.

- As a field of study, it deals with complex, rich, polymorphous, multidirectional relations among 'mediating conjunctures'.[19]

- As a dynamic, intermediality is what makes the production and evolution of such 'mediating conjunctures' possible; it also leaves in its wake media residues, scraps of uncom-

---

16   Jacques Rancière, *Le spectateur émancipé*, La Fabrique éditions, 2008.

17   Daniel Sibony, *Entre-deux: l'origine en partage*, Paris: Editions du Seuil, 2003.

18   Gilles Deleuze, *Cinéma*, Paris: Éditions de Minuit, 1983.

19   For reasons to be explained later, we prefer the expression 'mediating conjuncture' to 'media'. During intermediality's 'mediatic period', media was the preferred term cf Éric Méchoulan, 'Intermédialités: le temps des illusions perdues', *Intermédialités: Histoire et théorie des arts, des lettres et des techniques / Intermediality: History and Theory of the Arts, Literature and Technologies*, 1, 2003, 9-27 https://doi.org/ https://doi.org/10.7202/1005442ar.

pleted mediation processes, the remnants of 'mediating conjunctures' that have passed into history.

• Consequently, we call for an original approach that can yield a better understanding of this field and this dynamic.

For the first intermedialists, it was vitally important to grasp media (to use the terminology of the day) as part of a larger dynamic from which they drew their power. From the outset, therefore, the producers and users of media played a central role in intermedial reflection, to the same extent as the materiality of mediation processes. The early intermedialists, particularly those interested in electrical media, brought new and sustained attention to bear on the social environment from which media emerge and on which they act. This interest in sociality is worth noting because the first intermedialists were often wrongly accused of technocentrism, when in fact they held that media practices could not be dissociated from the individuals who were their agents. The socialities associated with media or created by media – from the first telephones to contemporary social networks – are not comparable to traditional social classes nor to stable communities; hence the neologism 'sociomedialities' is used to refer to these socialities that are specific to media and shaped by the processes of naturalization and universalization that traverse media.[20]

Etymologically, the term intermediality derives from 'intermedia' which, as its morphology suggests, refers to what is located between media. The word intermedia was first used in the American neo-avant-garde in the 1960s. Dick Higgins, co-founder of the multidisciplinary artistic movement Fluxus, is credited with coining the term. Higgins was a poet and composer who was heavily influenced by John Cage. Through a series of experiments that decompartmentalized disciplines, media, techniques and technologies, and effected transfers between them, Higgins and his Fluxus confederates produced works that would now be described as intermedial, since they harnessed multiple media (or arts) – which is characteristic of multi-mediality – for purposes of artistic production, presentation and experience (when the work involved user experience). Their approach was not entirely new, insofar as its creative logic was consistent with that of the historical avant-garde and rested on a thoroughly conventional conception of media. They understood a medium to be a discrete entity, one that is identifiable, separable and relatively stable and independent, as were the dominant electronic media of the time (radio, television, cinema). In that 'pre-intermedial' age, media were perceived as monads, defined sometimes on the basis of their real-world effects, sometimes on the basis of their structure and workings.

Friedrich Kittler,[21] Frédéric Barbier and Catherine Bertho-Lavenir[22] subscribe to the first approach: they define media by what they do. According to Kittler, 'a medium archives things, transmits things or processes things'.[23] For Barbier and Bertho-Lavenir, a medium is

---

20   We shall return to the question of sociomedialities in Chapter 4.
21   Friedrich Kittler, *Optical Media*, trans. by Anthony Enns, 1 edition, Cambridge: Polity, 2009.
22   Frédéric Barbier and Catherine Bertho-Lavenir, *Histoire des médias: de Diderot à Internet*, Paris: Colin, 1996.
23   Friedrich Kitler, quoted by Galloway, *The Interface Effect*, p. 13.

[a]ny system of communication that allows a society to perform any or all of three essential functions: conservation, remote communication of messages and knowledge, and the updating of cultural and political practices.[24]

These definitions make no mention of the material dimension of media or the role of the user in the process. Éliséo Véron proposes a structuralist approach that encompasses these factors:

A medium is [...] a constellation composed of a technology *PLUS* the social practices of production and appropriation of that technology, when there is public access to the messages (regardless of the conditions of access, which usually involve payment).[25]

Understood as the practical application of knowledge to a specific field and the resulting ability to perform tasks, technology includes artistic practices. While the arts are distinguished from 'other' media by certain qualities and functions that make them, according to Lars Elleström's term, 'qualified media', they are media nevertheless. The elements Véron gathers under the heading 'technology *PLUS* practices' roughly correspond to the general but highly complex concept of 'device', which is central to intermedial theory and which we shall discuss later.

Like media archeologists, intermedialists attach considerable importance to the question of practice and related protocols and values, which often account for a medium's success or lack thereof. Practice is inseparable from the concept of 'device'.

We should note here two fundamental aspects of intermedial thinking: the question of usage, which belongs to the sociomedial dimension of media, and the material side of all mediation, which belongs to the technical and technological dimension. These are also the two facets of a device in Foucault's sense. Lisa Gitelman's synthetic definition of media embraces these elements:

I define media as socially realized structures of communication, where structures include both technological forms and their associated protocols, and where communication is a cultural practice, a ritualized collocation of different peaple on the same mental map, sharing or engaged with popular ontologies of representation.[26]

While stressing the importance of sociomedialities, Gitelman's definition remains attached to the concept of structure inherited from communication theory, which, as we shall see, intermediality has gradually abandoned.

The word intermediality originally appeared in German *(Intermedialität)*. It is believed to have been used for the first time by media philosophers (e.g. Ole Hansen-Löve in 1983) and then

---

24   Barbier and Bertho-Lavenir, *Histoire des médias*, p. 5, our translation.
25   Eliséo Véron, 'De l'image sémiologique aux discursitivtés. Le temps d'une photo', *Hermès*, 13, 1994, 45 (p. 51) https://doi.org/10.4267/2042/15515, our translation.
26   Gitelman, *Always Already New*, p. 7.

spread to media studies, where it was popularized by the German scholar Jürgen E. Müller,[27] a dominant figure in the foundational period of intermediality, just as the word intermedia migrated from the artistic practice of the Fluxus group to the realm of media theory.

> At that period [the 80s], the isolating tendencies of media theories and histories and the rather banal fact that no medium could be considered as a 'monade' have motivated me and also some other scholars to direct our attention towards the intricate and complex processes of media interactions or media encounters. The notion of intermediality was based on the assumption that there are no pure media and that media would integrate structures, procedures, principles, concepts, questions of other media which have been developed in the history of Western media and would play with these elements.[28]

It was therefore in German media studies *(Medienwissenschaften)*, and not in the artistic avant-garde, that this significant theoretical advance was made, shifting media researchers' attention from media to relations between media.

Remediation, Central Concept in Intermediality's Mediatic Period

We have mentioned the original geographic dispersal of intermedial studies as one explanation for the arduous, still-incomplete unification of the field and its slow climb to recognition, but that was not the only problem. The very term intermediality causes confusion. It is marked, one might even say dated, by the historical circumstances under which it arose. The combination of the prefix 'inter' and the stem 'media' initially attracted derision. Therefore intermedialists are interested in what lies between that which is between: presumably, it can't be much! While the premise is accurate – intermedialists do indeed attend to what is between the in-between – the clearly false conclusion illustrates the widespread initial distrust in academic circles of a field that presented itself as an 'indiscipline'. It is true, however, that the question of the 'in-between' and how it was conceived by the intermedialists formed the basis for the emergence of intermedial theory.

The scepticism about this focus on the in-between may have been due in part to the confusion created by the prefix 'inter'. From the outset, intermediality fit naturally enough into a quasi-chronological series of terms that share its prefix: intertextuality, interdiscursivity, interdisciplinarity, intersubjectivity. There was therefore a tendency to situate it within a reassuring but largely false *telos*. But reading intermediality in terms of the 'inter-' logic inherited from semiotics and structuralism distorts its essence. That confusion was aggravated by the juxtaposition of 'inter-' with 'media', which duplicated its meaning. Some clarifications are in order here.

---

27   Jürgen Müller, *Intermedialität: Formen Moderner Kultureller Kommunikation*, Münster: Nodus
      Publikationen, 1996.
28   Müller, *Intermedialität*, p. 70.

In a study of the concept of 'medium'[29] – that which is at the center, in the middle – Jacques Rancière stresses the order of precedence implicit in the 'inter-' principle. It may seem obvious but Rancière underscores that for there to be an in-between, there must be two terms – two texts, two discourses, two disciplines, two subjects – engaged in a mutual relationship in space, in time, in logic, in short within a framework or system where an in-between can come between. The French mathematician, philosopher and psychiatrist Daniel Sibony had addressed the question several years earlier in *Entre-deux. L'origine du partage*,[30] an essay that may be considered a first attempt to theorize the notion of the in-between *(l'entre-deux)*, a category from which intermediality is distinguished but with which it was and remains associated.

Sibony situates the concept of the in-between as an extension and transcendence of the poststructuralist idea of difference: 'Until now, we have lived and thought under the sign of *difference*' but 'fast-moving facts and events [...] mean we can no longer be content with using difference as our reference point'.[31] Sibony argues that while the idea of difference is fundamentally valid, it is no longer adequate to reality. Difference has lost its relevance and must therefore be expanded to the in-between, an 'effective [...] operator'[32] that makes it possible to grasp the complexity of the phenomena at work in the in-between. Sibony calls those phenomena trials *(épreuves)*, for the in-between appears to him fraught with tensions and rifts.

> Perhaps the trials of the in-between come down to movements of varying degrees of richness in which *an identity attempts to piece itself back together again,* to resolve itself (while believing it is resolving with others), to stand up like a Harlequin costume in the circus of the world [...] But beyond the piecing together that the in-between actualizes, it achieves full strength when, in its great burgeoning, it appears as an *origin* figure [...] which assumes or implies something other than rehashing an immobile 'in-between'.[33]

We should note here in passing the nostalgic dimension of the origin figure (the lure of putting the pieces back together again), which also tinges discourse surrounding media and, in many respects, the work of researchers in the field. This nostalgia is doubly paradoxical. The advent of talking pictures, which was widely celebrated as a triumph of technology, only restored the original, pre-existing unity of sound and image as it is experienced in daily life (without technological mediation). Technology 'pieced together again' what it itself had rent asunder, since it had initially been incapable of reproducing sound and image synchronously. In short, and this is the second aspect of the paradox, the progress of media means that in the future we will be better off – as we were before! We will return to this paradox later; at this point, it should just be noted that the origin myth is not unrelated to the myth of authenticity.

---

29 Jacques Rancière, 'Ce que "medium" peut vouloir dire: l'exemple de la photographie', *Appareil*, 1, 2008 https://doi.org/10.4000/appareil.135.
30 Sibony, *Entre-deux.*
31 Sibony, *Entre-deux*, p. 9.
32 Sibony, *Entre-deux*, p. 12.
33 Sibony, *Entre-deux*, pp. 15-16.

This idea of precedence, with its whiff of nostalgia, is precisely what is problematic for inter-medialists. In fact, it gave rise to the first major epistemic break produced by intermediality and is what distinguishes it from the other 'inter-' variants listed above. That break relates to both the nature of the two poles that circumscribe the in-between and their prior nature. While the 'medium' as defined by Rancière is between, intermediality is interested in an in-between that lies between other in-betweens. In this sense, the detractors' gibes were quite on point. Yes, it's true, intermediality is interested in what lies between that which is between, and that is actually a great deal, immense, for there is nothing else: the two prior terms that bound Sibony's in-between are themselves between. In short, we have a between-two-betweens, and this is complicated by the fact that these in-betweens are dynamic, alive and constantly changing. The disappearance of the pre-existing, fixed poles that had been the starting point not only has a destabilizing effect but ultimately excludes any precedence and relegates the origin figure to mythology – and with it the nostalgia for an (idealized) past we can seek to recover.

The first intermedialists did not initially question precedence but – and this was already a sig-nificant shift – they reversed its elements, as Jens Schröter noted in 2012[34] in a retrospective discussion of the development of intermedial thinking.

> [W]e have to recognize that it is not individual media that are primal and then move toward each other intermedially, but that it is intermediality that is primal and that the clearly separated 'monomedia' are the result of purposeful and institutionally caused blockades, incisions, and mechanisms of exclusion.[35]

As can be seen, Schröter does not go so far as to question the origin figure defended by Sibony, nor the existence of media. He is content to invert the usual order – which places the two poles first, and then the in-between – and cast media as the product rather than the source of intermedial relations.

Before considering the reasons for abandoning the idea of precedence, we will look briefly at those 'blockades, incisions, and mechanisms of exclusion' that are at the heart of the intermedial dynamic and were the focus of researchers' attention during intermedial theory's mediatic period. Jay David Bolter and Richard Grusin aptly sum-marize this period with the punchy slogan 'a medium is that which remediates'.[36] Their book *Remediation: Understanding New Media*, published in 2000 some 15 years after the emergence of intermedial thinking, is a milestone in the brief history of intermedial theory that models the intermedial dynamic for the first time. It is also a pivotal work that takes stock of the progress of intermediality from the late 20th century to the dawn of the 21st and opens up new perspectives that are central to contemporary intermedial thinking. Bolter and Grusin thus offer a rich synthesis of mediatic intermediality while

34   *Travels in Intermedia[lity]: Reblurring the Boundaries*, ed. by Bernd Herzogenrath, Interfaces: Studies in Visual Culture, 1st [ed.], Hanover: Dartmouth College Press, 2012.
35   Herzogenrath, *Travels in Intermedia[lity]: Reblurring the Boundaries*, p. 30.
36   Bolter and Grusin, *Remediation: Understanding New Media*, p. 65.

at the same time noting its weaknesses and contradictions, heralding the next phase: postmediatic intermediality.

Basing themselves on Marshall McLuhan's maxim, picked up by Kittler and many others, that 'the "content" of any medium is always another medium',[37] Bolter and Grusin contend that

> [A medium] appropriates the techniques, forms and social significance of other media and attempts to rival or refashion them in the name of the real. A medium in our culture can never operate in isolation, because it must enter into relationships of respect and rivalry with other media.[38]

The remediation principle, which definitively excludes the possibility of monadic media, goes beyond the intuitions of the first intermedialists in that, while remaining focused on the development of media, Bolter and Grusin shun questions of precedence and origin. If a medium is something that remediates, then there can have been no first remediation. Lisa Gitelman encapsulates this disconcerting logic, which runs counter to tradition and to the Sibony / Rancière school of thought, in a phrase we have already quoted: a medium is 'always already new'. While it does not question the reality of media – on the contrary, that is its basis – Bolter and Grusin's model understands media as being in perpetual motion, perpetual remediation. Their emphasis is based squarely on what a medium produces and what it is produced by. In this sense, they substantiate Chiel Kattenbelt's suggestion that intermediality is currently the most radical form of performativity.

That being said, Bolter and Grusin's conception of media remains marked by the in-between, the medium being located between an independent reality and the representation conveyed by the medium, which we call the representational paradigm. For example, according to Bolter and Grusin, remediation operates 'in the name of the real'. This is a key point, for the relationship to reality that Bolter and Grusin defend has profound implications for various aspects of the intermedial dynamic. Ultimately, it sustains the notion that the medium, which both results from and produces remediation, remains a discretizable entity and its operation is limited to carrying, transforming or producing data, which is what Lars Elleström calls a 'media container'.

Bolter and Grusin's conceptual apparatus – starting with the concept of 'immediacy', a quality they ascribe to some forms of mediation – is built on this paradigm. By their definition, immediacy exists when mediation is imperceptible to the user, who may thus believe himself or herself to be in the presence of something which, in fact, is only being represented. Immediacy ties in with another concept, media 'transparency': the medium disappears, or to be more precise its devices and mediation processes are concealed. Here, Bolter and Grusin buy into the art of illusion, into a logic that returns us to mimesis and repeats the traditional arguments on which the media industry has constructed its marketing strategies for more than a century: the promise of 'fidelity' and 'high fidelity' (to what is unclear), touted by the

---

37 McLuhan, *Understanding Media*, p. 23.
38 Bolter and Grusin, *Remediation: Understanding New Media*, p. 65.

music recording industry in particular since the beginning of the 20th century. The visual reproduction industry's claims to steady progress hinge on the same illusionistic rationale. Bolter and Grusin's phrase 'in the name of the real' must be read in this light.

Though Bolter and Grusin remain deeply indebted to the mimetic model, they clearly are well aware that there are other forms of mediation that do not refer to a tangible, autonomous reality outside of, detached from, prior to and independent of the mediation process. Along-side the logic of immediacy, they develop the idea of 'hypermediacy'. But while the former is clearly explained, the argument supporting the second is laboured and incomplete, illustrating both its complexity and the difficulty Bolter and Grusin have in articulating it.

The concept of hypermediacy evokes the 'hypermedia' described by Ted Nelson in his famous 1965 article 'A File Structure for the Complex, the Changing and the Indeterminate'.[39] Nelson uses the term in conjunction with 'hypertext', a word he also coined. He pointed to the then-novel fact that the computer makes non-linear document management possible by hyperlinking content. That capability endowed media with an additional, vitally important dimension: they could now assume mutable structures that changed shape with every reading. To illustrate the point, Nelson cited the example of non-linear arrangements of film recordings that make them browsable in multiple dimensions, rather than the two dimensions of linear sequencing. Bolter and Grusin's hypermediacy is governed by the same logic.

The manifestations of hypermediacy are more diverse than those of immediacy and do not summon any external reality. To clarify this first fundamental aspect of the concept, it might be compared with an abstract painting, which in theory refers to nothing save itself and its own make-up (unlike a figurative painting, which despite its weighty artifice-creating apparatus, including perspective, produces an impression of reality). But hypermediacy is more than that. The hypermedial image Bolter and Grusin use to illustrate the idea is a page from the World Wide Web (unaccompanied by sound, in their example) that displays on the screen the system of hyperlinks described by Nelson. The page consists of disparate elements – letters, numbers, icons, photos, images of all kinds, fixed and moving – that share screen space within a system of complex, constantly changing and shifting relationships. Unlike the trans-parency effect which, as has been noted, works to suppress the perception of mediation, the webpage offers the spectacle of mediation at work. Against the immediacy-related principle of transparency, there is the principle of *opacity*. While transparency satisfies the user's need to experience the illusion of reality, opacity appeals to the user's fascination with mediation processes and the pleasure of seeing them in action.

> [T]he artist (or multimedia programmer or web designer) strives to make the viewer acknowledge the medium as a medium and to delight in that acknowledgement.[40]

---

39   T. H. Nelson, 'Complex Information Processing: A File Structure for the Complex, the Changing and the Indeterminate', in *Proceedings of the 1965 20th National Conference*, ACM '65, New York, NY, USA: ACM, 1965, pp. 84-100 https://doi.org/10.1145/800197.806036.

40   Bolter and Grusin, *Remediation: Understanding New Media*, pp. 41-42.

The webpage is a window that does not function as a metaphorical threshold, like the screen in traditional cinema or television, which once crossed admits the viewer into a reality beyond the screen. The webpage is a window that opens only onto other windows, never onto a three-dimensional reality that has or might have an independent existence. Richard Lanham sums up the difference with a change of preposition: *looking at* rather than *looking through*. Bolter and Grusin put it this way:

> [C]ontemporary hypermediacy offers a heteregoneous space, in which representation is conceived of not as a window on to the world, but rather as 'windowed' itself – with windows that open on to other representations or other media. The logic of hyper-mediacy multiplies the signs of mediation and in this way tries to reproduce the rich sensorium of human experience.[41]

The last remark foreshadows the major debates that have dominated intermedial theory since 2005, revolving around the gradual abandonment of the concept of medium in favor of mediation. In some artistic practices, such as the theater, this move has been paralleled by the shift from the representational paradigm to the presentational paradigm. These concepts are developed by Tracy C. Davis and Thomas Postlewait in connection with theatricality, the representational paradigm being governed by mimetic logic while the presentational paradigm obeys the logic of performance, in which one does not represent reality, but is in it.

Bolter and Grusin's essay marks the high point of this first period of intermedial theory, which we have called the mediatic period, and also reveals its limits, to which we shall now turn.

Postmediatic Intermediality: Remediation and Beyond

According to Alexander R. Galloway, the author of *The Interface Effect*, '[t]he remediation argument [...] is so full of holes that it is probably best to toss it wholesale'.[42] It is a harsh judgment but not without foundation: the remediation model is in fact seriously flawed, both in its theoretical underpinnings and its ability to account for the tremendous diversity of media interactions, which only increases as we refine our observational tools. It quickly became apparent that remediation, as conceived of and presented by Bolter and Grusin, describes only one type of interaction and cannot be considered a general scientific model of interme-diality, as its originators suggest. In a recent essay on intermedial dynamics, Lars Elleström uses the more cautious terms 'media transformation' and 'transfer of media characteristics among media' to better reflect the diversity which, he notes, is observable in analyses of both the development of media (the diachronic axis) and their modes of coexistence (the synchronic axis).[43]

---

41   Bolter and Grusin, *Remediation: Understanding New Media*, p. 34.
42   Galloway, *The Interface Effect*, p. 20.
43   Lars Elleström, *Media Transformation: The Transfer of Media Characteristics Among Media*, Houndsmill, Basingstoke, Hampshire New York: Palgrave Macmillan, 2014.

We do not share Galloway's view that the remediation model should simply be jettisoned. However, given its limitations, it does need to be enriched and made more complex, for Bolter and Grusin confined themselves to a fairly summary description of the mechanisms in play. Taking our cue from Elleström, we also need to situate the model within a broader framework, where it coexists with other types of media interactions that must also be described. In this chapter, we will discuss this expansion of the remediation model and look at some other types of interactions. But first, we will discuss some of the weaknesses that have been cited in the model's foundations, particularly the vein of essentialism and questions of origin and precedence.

Limitations of the Remediation Model

Intermediality rejects essentialist attempts to ascribe stable, irreducible properties to media. The essentializing temptation appears to us to be a remnant and legacy of the old debate over the paragon that has marked thinking about the arts since the Renaissance and led to their differentiation and hierarchization. The impulse to identify and classify the mediatic dynamics crystallized in media likely derives in large part from this paragonic reflex, triggered by the fact that the first great mediations of the electrical age appeared in the ontology-obsessed 19th and 20th centuries and possessed, or laid claim to, an artistic dimension.

We have outlined how intermediality deals with dynamics that consist of simultaneous movements of convergence, fragmentation, opposition, fusion, retreat and advance that may be slow or abrupt, predictable or chaotic. In this constant churn, nothing is fixed or assured. As we have noted, while the remediation model is quite useful for clarifying certain intermedial exchanges, its originators' attachment to the (essentialist) concept of medium, inherited from classic communication theory, limits its effectiveness and coherence. How can we reconcile their resolutely dynamic definition of media – 'a medium is that which remediates' – with traditional definitions that regard a medium as an identifiable, stable entity with relatively clear contours, that construct it as an object?

The origin figure is a key element in the essentialist arsenal, serving as sanction and performing a legitimating or delegitimating function, as the case may be. It is true that Bolter and Grusin's remediation model partially evades it by excluding any first remediation – there was therefore no original medium and their understanding of media is not monistic – but nevertheless their model remains attached, as we have seen, to the idea of precedence, through both the logic of the 'in-between' and the primacy they assign to the reproductive function of media, returning us to the general principle of mimesis in which remediation's central concepts of transparency and immediacy are grounded. Mimesis postulates the priority of reality to replica, the intelligible to the sensory, the immaterial to the material. It rests on the idea of a source or model, an idea as vague as it is problematic. But in both the priority of nature which inspires a work of art, and in that of past works which have influenced an artist, there is a common thread: the premise of precedence is colored by nostalgia. Curiously, the essentially electrical modern media technologies that proliferated from the 1880s, propelling their users into an unknown but promising future, also played on feelings of nostalgia and evoked, implicitly or explicitly, a return to paradise lost, the place where all was clear, pure and true. Leaving aside the religious dimension of this yearning, we will concentrate on its corollary, the trust in progress that permeates Bolter and Grusin's essay and has left its imprint on the thinking of the pioneers of intermediality as well as the media industry's discourse since the invention of electricity. The belief in progress makes nostalgia bearable, just as it nurtures it.

This attachment to the idea of progress explains why the first great intermedial investigations, extending up to the dawn of the millennium, dealt almost exclusively with 'successful remediations', i.e. with remediations that led to the creation of successful 'new' media (all electrical or digital, save a few exceptions such as the stethoscope, the gramophone and the first phonographs). Emblematic examples include records, cinema, radio, television and video. This focus stemmed from a Darwinian conception of the evolution of media. As the media industry's discourse suggests, media evolve just as living organisms do; we may therefore assume that they move from success to success in accordance with universal rules. Every step forward brings us closer to the absolute medium, the medium that will one day be capable of mediating everything with full transparency. We have already pointed to the fundamental paradox here: in the future, technology, including media technology, will make us still better off, as we were before. The absolute medium that will make this possible is the mythical black box, to which we shall return.

The optimistic, linear vision of media evolution that runs through Bolter and Grusin's essay and infuses the mediatic period of intermediality, which clearly rests on an old, essentialist substratum, is, however, contradicted by reality. Greater 'fidelity' to the 'source' – since media are replicating it with increasing accuracy – has been so vaunted by the media industry for more than a century because it revives the promise of a direct and true relationship to the external world. Here we clearly see the simplistic nature of the remediation model. If the obsessive quest for increased transparency – which often goes by names such as fidelity, high fidelity, high definition, ultra-high definition, stereo or surround sound – really were the sole driver of remediation, how can we explain the fact that low-resolution mediations like web documentaries shot with cellphones and MP3 tracks have not only survived, but continue to thrive? How are we to explain the popularity of grainy shots in cinema in an age of evermore powerful cameras?

While intermediality rejects essentialism, it also defies deterministic thinking. Media transformations are not governed by a single, predictable logic that is reducible to the idea of progress, as the remediation model implies. Not every remediation process leads to the emergence of a successful new medium. Remediation also creates losers, failures, scraps, all that is not 'selected' by evolution. Attempts at remediation can trigger unpredictable, chaotic reactions that do not follow the straight path of progress. This less glamorous aspect of remediation was largely ignored during the mediatic period of intermediality. It was not until the 2000s that researchers undertook a more thorough analysis of aborted remediation attempts and began considering media that, unlike theater for example, did not survive remediation(s) by cinema, radio or television, or survived in a weakened state, losing influence and *mainstream* status. This essential aspect of remediation, the survival of what Charles Acland has called 'residual media',[44] is entirely ignored by Bolter and Grusin. But the evidence clearly suggests

44   Charles R Acland, *Residual Media*, Minneapolis: University of Minnesota Press, 2007.

that cases of aborted remediation – remediation that does not yield a new medium – are the rule, not the exception.

The essentialist temptation and the attachment to the idea of progress are not the remediation model's only weaknesses. Though seductive, its underlying tetrad – immediacy / transparency, hypermediacy / opacity – is also problematic. As we have noted, for Bolter and Grusin, remediation operates 'in the name of the real', which is to say its goal is to produce the greatest possible transparency or, to use their terminology, to create the best possible immediacy effect (in rendering the object of representation). This first pair of the tetrad is opposed to the second – hypermediacy /opacity – within a simplistic, dualistic system that is presented as self-evident, when in fact the opposition is in no way preordained. To be sure, the tetradic structure is not really new: it derives from great debates in art history dating from the end of the 19th century about the nature and function of art, and more specifically the status of mimesis, a term of oft-noted polysemic import: 'imitation, copy, tracing, transcription, translation, execution, incarnation'.[45] In an important essay entitled *Transparence et opacité. Essai sur les fondements théoriques de l'art moderne,*[46] the art historian Philippe Junod, drawing heavily on the work of the German philosopher Konrad Fiedler (1841-1895), shows how the development of the machine in the 19th century paradoxically contributed to 'a revalorization of manual labour',[47] beginning with the work of artists. Their previously little-studied *praxis* began attracting the interest of analysts and of the artists themselves. The act of creation, particularly painting, becomes a legitimate object of study and

> the ferment that accompanies the testing of new techniques enters theoretical consciousness and leads to the foregrounding of the *opacity* of the materials.[48]

Later, Junod argues that opacity is what makes it possible to 'preserve the perceptibility of artifice'.[49]

> The theme of the value of artifice as a specific characteristic of art, which was picked up by romanticism and passed on to symbolism and then Russian formalism, was present throughout the [19th] century and its development led to the theories of *theatricality* [...], litterarity [...], pictoriality [...].[50]

In other words, to specific forms of mediation and mediality. In his demonstration, Junod aligns transparency with mimesis and opacity with *poiesis*. While the first principle, says Junod, dominated artistic practice for nearly two thousand years, the second is currently displacing it and informs modern art, in which the work is independent and functions autonomously, without transparency or referent, outside of mimesis.

---

45   Philippe Junod, *Transparence et opacité: Essai sur les fondements théoriques de l'art moderne. Pour une nouvelle lecture de Konrad Fiedler*, Nîmes: Chambon, 2004, pp. 11-12, our translation.
46   Junod, *Transparence et opacité.*
47   Junod, *Transparence et opacité*, p. 293.
48   Junod, *Transparence et opacité*, pp. 309-10.
49   Junod, *Transparence et opacité*, p. 393.
50   Junod, *Transparence et opacité*, pp. 310-11.

In Bolter and Grusin, Junod's equations

transparency = mimesis, opacity = * poiesis *

are replaced by

transparency = immediacy, opacity = hypermediacy.

These parallels should not surprise us; they betray the debt that the creators of the remediation model owe to art history and aesthetics. Nor is that influence new; it is evident in the first major works on electrical media.

There are two possible charges against Bolter and Grusin: that they hewed too closely to this model transplanted from another field, or not closely enough. The idea of *poiesis*, inherited from Fiedler, is dominated by the central principle of action and activity. Everything happens in and through action; all flows from it. This eliminates any precedence and, most importantly, makes performativity the basis of the creative dynamic, for both creator and user. It is surprising that while Bolter and Grusin note the existence of non-mimetic mediality, particularly in their discussion of hypermediacy, they do not delve more deeply into what is a vital dimension of intermediality. We shall return to this point in Chapter 3.

In this specific case, Bolter and Grusin fail to apply the artistic model strictly enough, but in others they would have done well to depart from it. Junod himself enriches the model's complexity with, for example, the concept of 'obscured transparency',[51] which he uses to account for works that proceed from mimetic logic but are ambiguous and assume meaning only through the action of the spectator/user. What Junod, and through him Fiedler, are stressing here is the sociomedial side of all mediating practices, as much as their performative dimension.

The absence of this consideration is acutely perceived in the passages in which Bolter and Grusin attempt to articulate immediacy, transparency and opacity. Either mediation is opaque, in which case there is no immediacy, or it is imperceptible, in which case there *is* immediacy. Aside from the fact that this is a simplistic alternative – as the category of 'obscured transparency' demonstrates, reality is more complex – Bolter and Grusin fail to take into account the user's agency and capacity to change and adapt. In her study of the spread of the telephone in the US, Carolyn Marvin traces its transition from curiosity, in the technology's 'magical moment', to part of everyday life as people assimilated its protocols, from learning how to manipulate the device to adopting a telephone voice.[52] This 'naturalization' of the device cannot be analyzed by simply wielding concepts of opacity and transparency for the simple reason that, at the end of the day, the device remains. We come to notice it less, or not at all, but this does not mean that it has become transparent, and that evolution does not produce

51   Junod, *Transparence et opacité: Essai sur les fondements théoriques de l'art moderne. Pour une nouvelle lecture de Konrad Fiedler*, p. 387.
52   Marvin, *When Old Technologies Were New: Thinking About Electric Communication in the Late Nineteenth Century*, pp. 63-108.

the effect of immediacy. We therefore need to revisit the basic equations that correlate opacity with hypermediality and transparency with immediacy, and enrich the model with new elements, such as the 'obscured transparency' category we have mentioned, inopacity, and so forth, to make it truly operational.

Enriching the Remediation Model:
Media Resistance, Demediation Opacification

We have seen the breakthroughs achieved by intermedial research up to the mid-2000s, and looked at resistance to remediation and the media residues left by remediation. This is fertile ground. Studies of sound in theater and still more recently of magic shows are examples of the new research paths that continue to improve our understanding of an intermedial reality composed of interlacings, intersections, 'entanglements', retreats and accidents, which the necessarily linear genealogy of media that inspired the first intermedialists could not readily account for.

Sibony linked the in-between to the origin, but also regarded it as a locus of trials, and it is true that remediating processes sometimes, though not always, produce reactions of varying degrees of violence governed by a general principle that we call 'media resistance', which may be understood as an attempt at counter-remediation. This too is entirely absent within the scope of Bolter and Grusin's analysis. Media resistance is a defence mechanism triggered when the basic elements of the mediating conjuncture are, or more precisely appear to be, jeopardized by the intrusion of an exogenous element. The resistance mechanism can have the effect of preventing, avoiding or delaying the introduction of the new element, or even expelling it after it has made an incursion, in which case we can speak of 'de-remediation' or simply 'demediation'. The decline in the use of fixed sound systems in some theaters with the advent of talking pictures is a striking example. Analyses of technical specs and prompt books show that 'technologically reproduced sound', which had made a widely noted and welcomed entrance to the theater stage, was gradually evicted from it. And the evidence clearly suggests that this expulsion was not related solely to the waning of its drawing power (as a curiosity or novelty).

To be sure, there is no essentialism at work here. When a 'medium' resists an incursion by a new technology (which is generally what the newcomer is), it is not behaving like a living organism secreting antibodies to expel or destroy an intruder. The resistance is always of a sociomedial nature: it arises in the community of men and women in which the medium operates; it springs from values; it is tied to a discourse, to representations. The microphone, which was long rejected by the theater – although today its use is becoming increasingly common, so much so that it goes unnoticed – was readily accepted on the cabaret stage for use by singers. And yet, since the days of Shakespeare, theater reviews had been complaining that it was hard to hear the actors. They did not articulate clearly enough or did not speak loudly enough; the audience was too noisy; or the sounds emanating from the stage or the building were too loud; or the acoustics were lousy. The arrival of sound systems could have solved those problems and enhanced theatrical mediation. But the theater resisted. Meanwhile, the cabaret halls put in sound systems and pulled in the crowds.

Media resistance is governed by a two-sided logic: it is a reaction to remediation, i.e. to the tendency of any media practice, artistic or not, to incorporate into its mediating action knowledge, techniques, technologies, values, behaviours and protocols that are foreign to it, and at the same time it checks the tendency of those elements to invade an established media practice. It must be said, however, that much less resistance is mounted against content than against production processes, as the rich tradition of adaptation (of novels for the stage, of plays for the movies, etc.) shows.

Media resistance also points to the fact that remediation doesn't only happen between media, it can also strike elements – again, often technologies – that are not attached to any specific medium, as was seen during the 'great digital conversion' of the late 20th century. Nor is it a one-way process. What we can extract from this is that media are not the only things that remediate or are remediated, and remediation produces all manner of effects, including media resistance and demediation (also illustrated, in a way, by the renewed popularity of vinyl records and the preservation of 16 mm and 35 mm film).

One of those effects, a somewhat surprising one, relates to the question of opacity. According to Bolter and Grusin, opacity exists when mediation is perceptible to the user. But Bolter and Grusin never consider the origins of opacity. Let us take the example of the use of the microphone in the theater. For reasons we will discuss later, the theater, like the opera, initially banned the mic from the stage, while three major media practices – records, radio and talking pictures (to which we might add cabaret) – developed and triumphed because of it. The radio played a central role in this process. The radio microphone truly revolutionized the art of public speaking, sparking an explosion of vocal styles. With radio, says Jacob Smith, the audience discovered that private speech and everyday conversation, could be elevated to the level of public speech and reproduced as such (or at least so it was perceived) on the radio in its infinite diversity, without exaggeration, without the familiar stage effects, 'transparently'. It was the dawning of the age of the crooner, the dulcet-toned singers who, instead of making the back-row seats shudder, made hearts flutter. President-to-be Franklin Delano Roosevelt saw the change and grasped the persuasive power of the intimate new medium. He started his *fireside chats* In 1929, at the beginning of the Great Depression, and continued them until 1944, into his third term as president. In those radio addresses, he talked to his fellow Americans about the great problems of the day in conversational terms. Radio sounded true. And the audience could tell the difference. The comparison with theater, another locus of public speech, was inevitable. As a result, observes Smith, 'vocal conventions of the histrionic stage were becoming all too audible as conventions to many listeners'.[53]

The new aurality created by radio brought out the artificiality of the voice and diction of stage actors, which had always been based on projection, volume and eloquence. Those fundamental qualities had been so assimilated by users that they no longer noticed them. They had become 'naturalized'. Until the end of the 19th century, i.e. the advent of sound reproduction

---

53 Jacob Smith and American Council of Learned Societies, *Vocal Tracks: Performance and Sound Media*, Berkeley: University of California Press, 2008, p. 94 https://login.gbcprx01.georgebrown.ca/login?url=https://ebookcentral.proquest.com/lib/georgebrown-ebooks/detail.action?docID=358943.

technologies, a fine voice was necessarily a powerful voice. The theater decided to stick with that. The contrast with radio voices laid bare the incongruity of theatrical practices, which used the pretext of 'truth', meaning unmediated, supposedly 'natural' relationships between actors and spectators, to celebrate an unnatural way of speaking.

Understood in terms of the logic of remediation, this development made opaque an aspect of mediation that had previously been naturalized (the mediation having become partially or totally transparent). This process is called 'opacification'. In our example, one medium, radio, rendered the mediation effected by another, theater, opaque. Thus, contrary to what Bolter and Grusin suggest, opacity does not spring solely from a medium's bid to assert its identity or from a developmental dynamic, but can arise from a conjuncture on which it may have no purchase. When opacification is not produced by the medium itself, it is almost always hostile to the medium and damages its image.

The mechanisms of media resistance, demediation and opacification do not exhaust the possibilities for further development of the remediation model, but the model's scope still remains limited, and it is clear today that this product of an age obsessed with recycling – industrial, esthetic, cultural, philosophical – cannot by itself account for the complexity and diversity of the intermedial dynamic. Other types of media interactions have been uncovered recently. In the following section, we look at the concepts of hypermediality, convergence and transmediality, which for the past 10 years have been at the center of intermedial reflection.

Hypermediality, Convergence and Transmediality: Other Models of Intermedial Dynamics

The concept of hypermediality emerged in the field of theater studies in the mid-2000s. From the intermedial point of view, theater is, of course, a medium. Indeed, it is the quintessential intermedial practice insofar as it is itself made up of interacting media – music, set design, plastic arts, dance, gesture, the actor's performance, the script, etc. – just as a webpage is composed of interacting elements. To put it in Bernd Herzogenrath's terms, theater is an *intermedium*, in Jens Schröter's terminology, it is a *super medium*. Chiel Kattenbelt goes still further, arguing that theater is a hypermedium but in a rather different sense than what Ted Nelson means by the term, as the following passage makes clear:

> [F]ilm, television and digital video are technology-based media that can record and play back everything that is visible and audible, within their specific ranges of sensitivity, but they cannot incorporate other media without transforming them under the conditions of the specificity of their own mediality. At the very most, media can *remediate* (Bolter and Grusin 1999) other media, which implies in the end a refashioning. Clearly, theatre is not a medium in the way that film, television and digital video are media. However, although theatre cannot record in the same way as the other media, just as it can incorporate the other arts, so it can incorporate all media into its performance space. It is in this capacity that I regard theatre as a hypermedium.[54]

---

54   Chiel Kattenbelt, 'Theatre as the Art of the Performer and the Stage of Intermediality', in *Intermediality*

A 'hypermedium' must therefore be understood as a federating medium, not an assimilating or transformative one. It is an 'archi-medium' or 'medium of media', *home to all*, an interart, a technological meeting place 'where the art forms of theater, opera and dance meet, interact and integrate with the media of cinema, television, video and the new technologies'.[55] This welcoming attitude is reflected in the theater's singular treatment of other mediatic objects, technologies, arts and media that enter into the process of theatrical creation or performance. The concept of hypermediality clearly derives from Bolter and Grusin's concept of hypermediacy, but it goes further. One attractive feature is that it eliminates the question of representation – and the relationship to reality – for, to return to Lanham's 'window' metaphor, it does not require us to choose between 'looking at' and 'looking through'. Better still, it allows for both simultaneously, which fits well with current theatrical practices and the performative, experiential approach. The concept of hypermediality not only enables Kattenbelt to underscore the singularity of theater compared with many other media, but also advances intermedial theory by revealing the existence of systems of mediatic and technological interactions that do not fall within the purview of Bolter and Grusin's remediating logic and do not partake in its cannibalistic violence.

In 2006, the same year that Chiel Kattenbelt developed the concept of hypermediality, Henry Jenkins tackled a central notion in the genealogy of media, one that underpins the concept of remediation: media convergence. It is not a new idea; its famous first appearance was in Wagner's call for a *Gesamtkunstwerk* – a total artwork – but it truly and lastingly took hold with the spread of electrical media in the first three decades of the 20th century. The advent of talking pictures is one of its most famous and memorable exempla, and also the most compelling argument in favor of the existence of a convergence dynamic. The talkies were in fact the result of the long-predicted 'convergence' of sound and image reproduction technologies, which until then had developed separately. Paradoxically, the forward-looking logic of convergence that Jenkins attacks has a kinship with the nostalgic quest for an imagined, pre-technological origin that has been shattered and whose fragments, to use Sibony's image, must be pieced together again. That convergence is supposedly driven by three main factors: continuous progress in the relevant technologies; technological determinism, which tends towards their fusion; and users' obsessive quest for absolute transparency. The convergence principle, which is held to be operative in all media dynamics, purportedly works 'in the name of the real', as the series of industry slogans we have already listed – fidelity, high fidelity, stereo, high definition, 3D, ultra-high definition, and so forth – assert. The relentless drumbeat of these claims over the space of more than a century constitutes a series of milestones – since they form a sequence – on the triumphant road that will ultimately lead to the absolute medium and perfectly transparent mediation, in Bolter and Grusin's sense. 'Sooner or later, the argument goes, all media content is going to flow through a single black box into our living rooms (or, in the mobile scenario, through black boxes we carry around

in Theatre and Performance, ed. by Freda Chapple and Chiel Kattenbelt, Amsterdam; New York: Rodopi, 2006, p 37.
55   Freda Chapple and others, *Intermediality in Theatre and Performance*, Amsterdam; New York: Rodopi, 2006, p 24.

with us everywhere we go)'.[56] Media convergence, then, is posited as the ultimate form of remediation, 'total remediation'.

Unlike Kattenbelt's hypermedia, which is federative and preserves the mediality of each medium, the black box is integrative, mixing and blending all that enters into it. The idea of convergence is so widespread and so frequently invoked by the industry that it seems natural and inevitable, but Jenkins demurs. In his view, technological progress cannot be separated from the conjuncture in which it arose and technological determinism is an illusion, a marketing ploy: 'convergence refers to a process, not an endpoint'.[57] Jenkins argues that '[p]art of what makes the black box concept a fallacy is that it reduces media change to technological change and strips aside the cultural levels'.[58] But '[c]onvergence does not occur through media appliances, however sophisticated they may become. Convergence occurs within the brains of individual consumers and through their social interactions with others'.[59]

One of Jenkins' great merits is that he has refocused intermedial reflection on the key role of the user (he uses the word 'consumer') and more generally on the ever-changing environment from which media emerge and in which they operate. He thereby converts the traditional concept of media convergence – spontaneously understood as 'technological' convergence – into a theory of comprehensive convergence,

> where old and new media collide, where grassroots and corporate media intersect, where the power of the media producer and the power of the media consumer interact in unpredictable ways.[60]

How then can we subscribe to any kind of teleology? The answer is simple: the black box is a myth created out of whole cloth by the media industry to justify planned obsolescence and its ever-faster product replacement cycle. But the current participatory culture, which is tied to the shift to performativity and embraces networking and collective intelligence, has other demands: users do not want, if they ever did, a 'one-size-fits-all relationship to media content. Consumers want the media they want, where they want it, when they want it and in whatever format they want'.[61] Media convergence supposedly led towards a single super-medium, the black box; this new participatory culture, which according to Jenkins is creating a 'convergence culture', leads in the opposite direction, towards media divergence. To adapt to new user behaviours, the industry needs to multiply its platforms, specialize them for specific purposes, and find ways to circulate its media content on them. This new dynamic set the stage for a decisive phase in intermedial thinking: the birth of the concept of transmediality, which illuminates intermedial relationships quite different from Bolter and Grusin's remediation or Kattenbelt's hypermediality.

---

56 Henry Jenkins, *Convergence Culture: Where Old and New Media Collide*, Revised edition, New York: NYU Press, 2008, p. 14.
57 Jenkins, *Convergence Culture*, p. 16.
58 Jenkins, *Convergence Culture*, p. 15.
59 Jenkins, *Convergence Culture*, p. 3.
60 Jenkins, *Convergence Culture*, p. 2.
61 Jenkins, *Convergence Culture*.

In the third chapter of his essay on the *Matrix phenomenon*, Jenkins shows that even as the industry continued celebrating steady progress towards the 'black box', it submitted to the logic of divergence by creating 'transmedial' media content, i.e. content that can be adapted to the various existing media technologies. *The Matrix* spawned three feature films, a 90-minute television series, animated films, multiple comic books, video games, etc.

Transmediality takes infinitely variegated forms, ranging from simply distributing purportedly uniform content on different platforms, which was the usual practice at the outset, to creating fragmented collective intelligence, whose different parts are supported by different technologies and the whole, like Paolo Soleri's arcology,[62] cannot be grasped from any vantage point.

In the first case, the objective is to adapt media content to different medialities (or media) with the intention of preserving the content in its entirety as far as possible. This conception of transmediality rests on two major assumptions: that media content can exist independently of mediation processes and, conversely, that those processes have no content.

In the second case, the semantic content is broken up according to the mediating qualities of the available platforms and the 'fragments' are distributed across the platforms. Thus, to return to the *Matrix* example, the user discovers new media content on each platform, since each supports an aspect of *The Matrix* that corresponds to its specific mediality. In practical terms, this means that no single user can master the entirety of the media content; only the sum of users has that grasp, and this relates to what is called collective intelligence.

There are some major issues here. Picking up on Jenkins' discussion, Mark J.P. Wolf analyzes the impact of transmediality on traditional distinctions between real and mediatized worlds.

> Transmediality implies a kind of independence of its object: the more media windows we experience a world through, the less reliant that world is on the peculiarities of any one medium for its existence. Thus, transmediality also suggests the potential for the continuance of a world in multiple instances and registers; and the more we see and hear of a transmedial world, the greater is the illusion of ontological weight that it has, and experiencing the world becomes more like the mediated experience of the Primary World.[63]

While transmediality does not totally blur the dividing line between what Wolf, referencing Tolkien, calls the primary world and the mediatized world, it makes the distinction superfluous; from the point of view of a transmedial logic in which user experience is paramount, the question of representation becomes moot, as does the interplay between immediacy and hypermediacy. As in the case of convergence, the emphasis is on the user and the user experience, which is always real. We have moved from the representational paradigm to the experiential paradigm.

---

62   Paolo Soleri, *Arcologie: la ville à l'image de l'homme*, Roquevaire: Parenthèses, 1980.
63   Mark J. P Wolf, *Building Imaginary Worlds: The Theory and History of Subcreation*, New York: Routledge, New York, 2012, p. 247.

Shattering the Medium: The Advent of Postmediatic Intermediality

Transmediality has had another notable impact on intermedial theory: it challenges the very existence of media as they were conceived of by the first intermedialists. In his definition of the medium, Véron did not content himself with underlining the technological dimension; he made technology the foundational element of a medium: 'a constellation composed of a technology *PLUS* the social practices of production and appropriation of that technology'. The 'double birth' model that André Gaudreault and Philippe Marion extract from their observations on the origins of cinema rests on this same principle of primacy. According to Gaudreault and Marion, the double birth of a medium unfolds in three phases: the appearance of a new technology (phase 1), the creation of a device (phase 2), and the realization of the medium's expressive autonomy and specificity, with all that this implies in terms of both socialities and the medium's institutional dimension (phase 3). François Odin describes the same process, calling the emerging technology a 'cryptomedium' (since it cannot yet be predicted that it will give birth to a medium), phase 2 being the appearance of a 'protomedium', and phase 3 the appearance of the medium as such.

Transmediality doesn't just disrupt this chronology; it denies the primacy of technology. Jenkins' critique of convergence culture, his defence of media divergence, and all that this implies for transmedial dynamics support the conclusion that a change in technology does not produce a change in medium. This also has consequences for the origins of media, as Jenkins makes clear:

> [H]istory teaches us that old media never die – and they don't even necessarily fade away. What dies are simply the tools we use to access media content – the 8-Track, the Beta tape.[64]

Jenkins ends up drawing a clear distinction between the medium and what he calls the 'delivery technology' for media content. 'Recorded sound is the medium. Cds, MP3 files, and 8-track cassettes are delivery technologies'.[65]

Lars Elleström's work in recent years on multimodality moves in the same direction. He divides a medium into three subcategories: the basic medium, the qualified medium, to which artistic practices are attached, and the technical medium, which corresponds to Jenkins' 'delivery technologies'.

> Basic and qualified media are abstract categories that help us understand how media types are formed by very different sorts of qualities, whereas technical media are the very tangible devices needed to materialize instances of media types.[66]

---

64   Jenkins, *Convergence Culture*, p. 13.
65   Jenkins, *Convergence Culture*, p. 13.
66   Lars Elleström, *Media Borders, Multimodality and Intermediality*, Basingstoke [u.a.: Palgrave Macmillan, 2010, p. 12.

We can see at once the ontological and methodological issues presented by this conceptual move and the concomitant risk of a reversion to an essentialist vision of media. To date, a medium's identity has been defined primarily by the material dimension of the mediation, but what happens if an important element of the mediation – Elleström's technical medium or Jenkins' delivery technology – is radically changed? Will this lead to the end of the medium? That is precisely the question at the center of Gaudreault and Marion's essay, *The End of Cinema? A Medium in Crisis in the Digital Age*. They ask whether 'cinema, in its shift to the digital, has simply made a turn (in which case we could speak of a digital turn) or whether it is in the process of becoming something else – whether it is undergoing a true mutation (in which case we could speak of a digital mutation)'.[67] The latter case will presumably need a new name. The authors quote, in passing, Raymond Bellour to the effect that 'digital technology is not enough to kill cinema. Although it may disturb it in many respects, it does not touch its essence'[68] and draw on the genealogy of cinema arguing that the movies are not dying; it is just that 'a generation stands aside and makes room for the next'.[69]

The genealogical argument is attractive. To the double birth model they developed in 2000, the second birth being the institutionalization of the medium, Gaudreault and Marion add a third birth, which is currently in progress: 'an integrative and intermedial birth involving a degree of return to porosity, to a hodgepodge, to hybridity, to crossfertilization - all of which imbued the medium's very first birth'.[70] By comparing the current state of affairs with an older period in the history of the medium, Gaudreault and Marion endow it with ontological legitimacy: digital is mediatically acceptable since the change it produces echoes a situation that is already inscribed in the history of the medium. They accredit the rupture thesis defended by Jenkins and Elleström: even a media practice as technologized as cinema can survive the most radical technological transformations.

Current intermedial thinking, i.e. that of the postmediatic period, is characterized by the erosion of belief in determinism and progress – and particularly the death of the illusion of mediating exclusivity – as much as it is by the shattering of the traditional conception of media and the demise of technological primacy. The question of a medium's identity has become secondary to its action and its position in transmedial configurations. Not only do we now know that the set of characteristics once used to define a medium is in fact unstable and in constant flux, but we have also understood that no characteristic is the exclusive property of any one medium and none, taken in isolation, is decisive for the medium's survival, not even an array such as the camera-projector-screen triad that appears so essential and basic.

The conclusion to which Gaudreault and Marion come doesn't just confirm Jenkins and Elleström's hypothesis about delivery technologies, since this applies to all components of

---

67   André Gaudreault and Philippe Marion, *The End of Cinema? A Medium in Crisis in the Digital Age*, New York: Columbia University Press, 2015, p. 6.
68   Raymond Bellour, *La querelle des dispositifs. Cinéma, installations, expositions*, coll. 'Trafic', Paris, P.O.L., 2012, p. 16 quoted by Gaudreault and Marion, *The End of Cinema? A Medium in Crisis in the Digital Age*, p. 249.
69   Gaudreault and Marion, *The End of Cinema?*, p. 183.
70   Gaudreault and Marion, *The End of Cinema?*, p. 123.

media; it spells the end of the traditional conception of a medium as summarized by Véron. Hence Alexander R. Galloway's dictum: 'not media but mediation'.[71] For Galloway, any attempt to find or impose a clear and stable definition of media is not only vain, but a pipe dream we must shake off, along with the technocentric thinking behind it. In a shift that reflects the ascendancy of the performative perspective in the field of media studies, mediation is now granted priority.

As Müller observes, intermedial thinking was born of a radical rejection, a break with the understanding of a medium as a 'monad' operating independently and analyzable in isolation. Proceeding from Marshall McLuhan's idea that every medium includes another medium, Müller and the first intermedialists showed that a medium is always in interaction with other media. They called this interactive dynamic 'intermediality'. They soon came to the realization that intermediality came first; it is intermediality that produces media. Revolutionary though it was, this first period of intermedial thinking remained focused on the concept of medium until the mid-2000s. Hence our label 'mediatic intermediality' for the first phase of intermedial thinking, which spanned the 20-year period from 1985 to 2005.

The difficulty of defining what a medium is, and the basic technological mutations that have made it impossible to confidently associate a medium with a given technology or technologies, have led researchers in intermediality to abandon the historical concept of medium in favor of the idea of mediation, or what we call 'mediating conjunctures'. Thus, over the past 10 years, we have gradually entered the postmediatic phase of intermediality.

---

71   Galloway, *The Interface Effect*, p. 36.

ESSENCE AND PERFORMATIVITY

Introduction

In Chapter 1, we described the history of intermedial thinking as a gradual transition from a theory of media, constructed as objects, to a theory of the end of media – a 'postmediatic' theory. This development was accompanied by a steady decline in essentialist ways of thinking about media. Intermediality can therefore be said to have sprung from the shift from an essentialist heuristic (based on the assumed existence of competing media) towards a non-essentialist heuristic (there are no media but rather dynamic convergences that produce forms of mediation, what we call 'mediating conjunctures' – see Chapter 3). We therefore divide intermedial thinking into two periods: the mediatic and the postmediatic.

We believe this important development calls for further analysis for two reasons. First, because it is a paradigm shift that cannot be understood as a linear progression, as a teleologically ordained advance from a flawed theory to a better one. In fact, both the essentialist and non-essentialist paradigms run through the history of Western thought. Indeed, the debate between their proponents may be considered a pivotal theme in the history of philosophy since its inception. Secondly, while essentialist thinking poses serious problems for our apprehension of the world, a radically and exclusively non-essentialist approach may land us in different, but equally insurmountable, theoretical pitfalls.

We might put it this way: the need to avoid reducing media practices to reiterations of actions that were predetermined by a medium's specificity calls for postmediatic thinking, but can we escape the need to characterize media practices and understand the specific features by which they are distinguished? In short, while the complex concept of 'medium', as understood prior to the postmediatic period, exposes the difficulties of a radically essentialist stance, its critique brings us up against the limitations of any non-essentialist position. If we try to define one medium's characteristics in order to distinguish it from another – to articulate the difference between a printed book and an e-book, for example, or between a play and a movie that tell the same story – we take medium and content to be mutually independent (the position defended by Jenkins[72] and Elleström[73]), which confronts us with the paradox of imagining content without a medium, ready to be shaped in one manner or another, depending on the chosen medium. If, on the other hand, we take the non-essentialist view and suppose that content and medium are inseparable, how then can we identify the characteristic features of a given medium, or even discuss them? How can we contemplate the very existence of media specificities? The non-essentialist perspective could logically lead to a sweeping conflation of media practices. In this case, ontological interrogations must be set aside, for media constellations do exist, if not in reality at least discursively. It would be naïve to think that non-essentialism could simply displace and replace essentialism.

---

72   Henry Jenkins, *Convergence Culture: Where Old and New Media Collide*, Revised edition, New York: NYU Press, 2008, p. 13.
73   Lars Elleström, *Media Transformation: The Transfer of Media Characteristics Among Media*, Houndsmill, Basingstoke, Hampshire New York: Palgrave Macmillan, 2014, p. 47.

Given the limitations of both positions, we will begin by expanding our critical analysis of essentialism from a philosophical perspective, drawing on the concept of performativity. In chapters 3 and 4, we will look at how the essentialist / non-essentialist contradiction can be transcended in order to escape the opposed perils of conflation and static categorization.

Essence and Flux: The Irreducibility of Movement

The history of Western philosophy starts with the apparently irresolvable aporia produced by the opposition between the one and the many. The first philosophers picked one side or the other, defending either the unity of reality or its plurality. Most of the questions that have claimed the attention of philosophers in the West and beyond arise from this opposition. Briefly put, unity is the notion that makes the ideas of essence and concept possible, whereas multiplicity grounds the possibility of motion and freedom. This debate remains pertinent to this day and the theoretical issues it throws up will inform our reflections on media.

In Book 1 of the *Metaphysics*, Aristotle discusses the ideas of Parmenides, the founder of the Eleatic School, who held that reality is one. As Aristotle notes, the difficulty with this position is how to account for motion, since Parmenides is 'compelled to accord with phenomena'.[74] As we observe the world, we see that it is in motion, that things mutate and change, that they do not abide in immobile oneness. The Eleatics maintained that observation of phenomena was deceptive: in reality, there is no motion and things do stay as they are. In his famous paradoxes,[75] Parmenides' pupil Zeno tries to illustrate the logical absurdities that ensue from the idea of motion and its incompatibility with reason. Motion is irrational. In the paradox of Achilles and the Tortoise, Zeno imagines Achilles racing a tortoise. Achilles, the much swifter runner, gives the tortoise a 10-metre head start. The race begins and Achilles quickly covers the first 10 metres. But in the meantime, the tortoise has also advanced a few centimetres. So it is still ahead. It will take Achilles only an instant to cover those centimetres, but in that time the tortoise will have moved a miniscule distance, which will still separate it from Achilles. Ultimately, Achilles will never be able to overtake the tortoise. From this line of reasoning, Zeno deduces that motion is an illusion: we inhabit a shifting world of appearances but in truth, all is immobile, for motion is a contradiction.

Other philosophers, of course, defend a diametrically opposed position: Heraclitus sides with motion and arrives at the conclusion that there is no unity or essence, only flux: 'No man ever steps in the same river twice',[76] for everything is in movement, 'everything flows'.

The opposition between the one and the many is central to the dialogue Plato entitled the *Parmenides*, believed to be one of his last. It has been read as self-criticism, for Plato seems

---

74   Described by Aristotle, *The Metaphysics*, trans. by Hugh Tredennick, Cambridge, Mass.: Harvard University Press, 1989.
75   Aristotle, *Physics*, ed. by C. D. C Reeve, 2018, p. 239b05 ff.
76   Hermann Diels, Walther Kranz, and Kathleen Freeman, *The Presocratic Writings.*, 2017, p. 22B12 http://pm.nlx.com/xtf/view?docId=presocratics_gr/presocratics_gr.00.xml [accessed 18 February 2019].

to be concerned with flaws in the essentialist view of the world, the very approach he had embraced with his theory of forms.

Plato poses a simple question: how can we move from one point to another? In other words, how are we to explain change?[77] The difficulty resides in the fact that any essentialist theory rests on the principle of non-contradiction. Nothing can have two contradictory properties: it is not possible for a thing (A) to have a property (+p) and not have it (-p). This principle is sometimes bracketed with the 'identity principle': the proposition that we cannot say a thing possesses an attribute and its opposite (p and not p) is the equivalent of saying that it is what it is.

Non-contradiction is therefore also the logical principle that allows us to ascribe a stable identity to a thing, to establish its essence. For a thing to have an essence, it must have certain properties and not their opposites. For example, according to the principle of non-contradiction, a chair cannot be white and, at the same time, not white. If we paint it blue, a change has occurred. The first chair is no more, and we now have a different chair. This is not a problem if we consider the two chairs, white and blue, separately. The difficulty arises when we try to establish a relationship between them. How do we get from the white chair to the blue one? What is in between the two? We would have to be able to speak of a chair that is white and not white, or blue and not blue.

This is exactly the paradoxical question that Plato asks in the *Parmenides*: how can we get from stasis to motion? Untangling the problem rationally poses daunting challenges: at a given point in time (T), there is a motionless object; at another (T_1), an object in motion. How can we account for this change? As can be seen, this is the same conundrum as the problem of the chair. First (T) there is a white chair; then (T_1) a blue chair. To progress from time T to time T_1, we must imagine a transition between the two, a paradoxical instant in which the thing is in motion and not in motion, or the chair is white and not white. In that instant, the principle of non-contradiction does not apply.

Plato calls this paradoxical transition by the Greek word *exaiphnes*,[78] which means, precisely, 'instant'. The *exaiphnes* is an in-between, an inter-space where opposites coexist; it might therefore be translated as 'interval'. Plato's discussion of the concept is highly complex and has been the subject of diverse interpretations for centuries. These may be organized into two opposed camps:[79] the *exaiphnes*, as the in-between, is either the locus of contradiction or a non-locus where mediation is effected. If the *exaiphnes* is a locus, it is a conflicted space, a place where opposites coexist, the tragic scene of a contradiction. This conception of the in-between is precisely what Sibony picks up on,[80] as we have seen. From this view, the *exaiphnes* is the site where the white chair and blue chair coexist. But the *exaiphnes* can

77 Plato, *Parmenides*, trans. by Benjamin Jowett, Adelaide: The University of Adelaide Library, 2008, p. 130e https://en.wikisource.org/wiki/Parmenides [accessed 21 December 2018] ff.

78 Plato, *Parmenides*, p. 156d.

79 For example, Fred Lawrence, *The Beginning and the Beyond: Papers from the Gadamer and Voegelin Conferences*, Chico: Calif.: Scholars Press, 1984, p. 7 ff.

80 Daniel Sibony, *Entre-deux: l'origine en partage*, Paris: Editions du Seuil, 2003.

also be conceived as a non-locus, something that escapes space and time, a void, an abyss. In the first interpretation, the *exaiphnes* is the instant in which a thing at once possesses a property and does not, in which the chair is white and not white. In the second, it is the mysterious nowhere in which the thing has neither one property nor the other, the chair is neither white nor not white.

The defining features of intermedial thinking are readily discernable in this logical structure: on the one hand, the attempt (typical of intermediality's mediatic period) to situate the in-between as the site where two or more media coexist; on the other, the jettisoning of the concept of medium in favor of media practices, proceeding from the concept of mediation. But in either interpretation, the *exaiphnes* poses serious problems for our attempts to rationally account for what we see. It is difficult to wrap our minds around any challenge to the principle of non-contradiction, for our thinking and language are predicated upon it. We shall return to this question in Chapter 4, where we examine processes of naturalization.

Discrete, Continuous, Digital

Our analysis of the philosophical premises underpinning the distinction between the mediatic and postmediatic periods in intermedial thinking can be taken one step further: the opposition between essence and motion can be correlated with another, more formal antithesis between discrete and continuous. As shall be seen, these two categories also ground the distinction between analog and digital, which is critically important in current discourses about media.

The discrete and the continuous are two opposed forms of representation of reality. The mathematical difficulty of rendering them commensurable is, in a sense, at the root of the paradoxes we have described. To define these two terms, which in general physics correspond to the distinction between particle and wave, it may be said that an object is discrete when it consists of a set of (finite or infinite) unitary, indivisible parts. Whole numbers are discrete: there is nothing between 1 and 2; the numbers 1, 2 and 3 are each unitary, indivisible elements. Similarly, a fruit basket is a set of discrete parts, for it consists of a series of indivisible units, each single fruit. By contrast, a continuous object is not made up of parts but is always further divisible, and its distinguishing property is density: between point A and point B there is always an infinite series of points. Real numbers are continuous. Water is continuous. One cannot have one water or two waters; it is a dense body.

The difficulty of understanding the transition from stasis to motion, and the paradoxes that attend interpretations of change in general, derive from the impossibility of explaining the continuity of reality starting from a series of discrete units. To return to our chair example, the observation that there was a white chair and then a blue chair clearly hinges on a discretization of reality: we isolate two objects and posit them as whole, unitary, indivisible parts of reality. White chair (1), blue chair (2). When we try to understand what happens between points 1 and 2, we are forced to shift from the discrete to the continuous, and this raises a number of difficulties, beginning with a mathematical dilemma.

The problem is that our language is inherently discrete (consisting of words, phonemes, etc.) while reality appears continuous. The attempt to understand the objects of perception using our discretizing apparatus is troubled by paradoxes, the resolution of which can be counted among humanity's loftiest aspirations. It might be said that all scientific thought – and technical thinking in particular – strives to refine our apparatus for discretizing reality in order to grasp the world as a continuous system without eliding its complexity.

Digital may be seen as the most recent and potent expression of this endeavour, which can be traced through the history of science from the Greeks to Leibniz to contemporary sampling theory. In its primary meaning, 'digital' refers to the recording of sounds and images (or other information) as discrete data points, as opposed to the continuous recording mode used by analog. Analog recording reproduces phenomena in a way that is analogous to reality, that is, continuous. The sound curve inscribed on a vinyl record is continuous; if we take any two points on it, we will always find an infinite number of points between them. The curve is therefore 'dense' in the mathematical sense of the word. By contrast, the digital principle breaks the continuity of sound into discrete parts. This discretization process is what is called 'sampling'. In practical terms, it consists in taking a continuous sound wave and selecting sample points from it. This method doesn't capture the sound as a whole but rather the changes between set intervals. A shorter interval will yield more precise sampling and therefore digital sound of higher quality. However, the sound has an inherently lower grade than that produced by analog recording, since it does not reflect the continuity of the original, but only a number – albeit a high number in many cases – of sample points. The discretization process simplifies the recorded sound by reducing it to a series of whole numbers, specifically 0's and 1's. This simplification makes the reproductions easier to manipulate.

To sum up, it can be said that analog is more 'faithful' to reality, for it is continuous, but it is very difficult to transport, transmit, archive, describe and manipulate, while digital is less 'faithful' since it is discrete but easier to work with. Ironically, the industry supported the move from analog to digital on the grounds that it would improve sound quality, repeating the logic of the original argument about fidelity and high-fidelity discussed in Chapter 1. This disconnect between discourse and reality is indicative of the complex, non-linear relationship between tools, practices and discourses that underpins sociomedialities, the constantly shifting civil groupings produced by social dynamics.

However, analysis of the digital / analog duality clarifies the main advantage of discretization. While it doesn't provide a 'faithful' representation of reality, it does yield a better purchase on reality, a way to master it, understand it, manage it. Discretization is the basis of technical thinking: if we want to have a working knowledge of the world, knowledge that enables us to act on reality, we must divide it into discrete parts.

Media and Action

We will now try to apply these formal principles to the question of media and mediation. The discrete / continuous opposition provides another angle from which we may attempt to chart a middle course between the radically essentialist and radically non-essentialist positions.

On the one hand, it is not possible to reduce the continuity of reality to a series of distinct, immobile units, which in our case would mean reducing media practices to stable media. On the other hand, we need to be able to articulate a (discretizing) discourse about them. How can we do so without identifying them, without marking their boundaries?

As we have seen, the concept of medium carries a heavy essentialist burden: the in-between is understood as a solid, a thing. In this case, media are objects; each medium – radio, television, the Web – has its own specificity. The essentialist approach renders shifting configurations discrete and graspable, but it thereby runs the risk of reifying them and suppresses a critical factor: action.

Defined in formal terms, an action is anything that can be expressed by a verb (other than a stative verb). An action can be of many sorts: loving, thinking, suffering, moving, feeling, and so forth. So an action exists where there is movement, a process. It is what occurs within the continuity and flow of reality. The idea of an in-between that serves as a passage between two discrete units therefore runs counter to the conception of action as a continuous movement that cannot be reduced to two poles. The irreducibility of action renders it incompatible with discretization. This is precisely what Zeno's paradoxes demonstrate: within the confines of a reality mapped as a series of discrete points, action is impossible. Achilles will never be able to overtake the tortoise. In an unconditionally essentialist frame of reference such as Zeno's, there is no room for verbs, with the sole exception of the verb 'to be'. The postmediatic approach therefore spurns the concept of medium in favor of mediation, a move that enables it to restore the centrality of action. This is the argument Alexander Galloway makes in his essay *The Interface Effect*, to which we shall return.[81]

In a way, the first intermedialists' efforts to understand the human and non-human mechanisms and agents, at work in the complexes they called media, served precisely to define media on the basis of stable, recognizable features, and therefore to essentialize them by focusing on their elements rather than on what they do, *i.e.* the process of mediation. This useful and indeed necessary fragmentation for the purposes of observation led the early intermedialists to define three structures of mediation, which they believed could be used to pin down the specificity of each medium: the *support*,[82] the interface and the apparatus. One medium could be distinguished from another by the particular type of mediation it performs. It was believed that this approach would make it possible to delineate the specific characteristics of media practices. In other words, by looking at the *support*, the interface or the apparatus, they would be able to grasp the specificity of the 'theatrical apparatus' in relation to the 'cinematic apparatus', or of a 'digital interface' compared with another interface, or of print as a *support* compared with a computer monitor. These three categories effectively encompass the normative dimension of media – that is, they articulate how a specific medium structures, shapes and determines an action. (As we shall see, these processes can also

---

81   Alexander R. Galloway, *The Interface Effect*, Cambridge, UK; Malden, MA: Polity, 2012.
82   Applied to media, the French word *support* refers to the physical material on which a medium rests and through which it operates, such as the printed page or a videotape. The concept of *support* is in some ways similar to the notion of 'delivery channel' or 'delivery technology'.

be described using Grusin's concept of premediation, or the idea of 'affordance' advanced by Donald Norman.[83]) The categories of *support*, interface and apparatus therefore perform a key function in differentiating media practices and grounding their identities. But, at the same time, they confront us with a series of aporias and paradoxes that cannot be ignored. After briefly defining the three terms, we will attempt to show why their use is problematic.

Support

Let us begin with the concept of *support*, a term with multiple meanings that pervades French discourse about media despite its polysemic character. Etymologically, it refers to something that is under something else and holds it up. It is a base that structures the thing it bears. This primary meaning can be illustrated by the holder or trellis that supports a plant, or by the verb 'to support' in the sense of providing physical or moral succour.[84] One can rest on a *support* or rest something on it. But when we look at a plant holder, we see that it isn't just a base on which a plant stands. A *support* is more than a foundation or a pedestal. Unlike a foundation, it structures and shapes the thing it holds up. A plant holder is the object on which the plant stands, to be sure, but it can also mould a plant, making it grow straight or at the desired angle.

We must begin by asking what it is the *support* holds up. Theoretically, it could be anything – a plant, a person, or (and this is what interests us) an idea, content or information. To speak of a *support* means there must be a thing that can be imagined – or at least named – that existed before it was supported. First there is a plant, and then, if it is placed against a trellis, it will grow as the trellis bends it; it will assume a given shape. First there are people, and providing them with support will help them live more fulfilling lives or at least ease their pain. First there are ideas, and the *support* helps them take form, materialize. This point bears repeating: speaking of a *support* presupposes the prior existence of the thing that will be supported. This idea is implicit in the term 'content': content is something that exists separately from the *support* that holds it.

Secondly, we need to ask how the *support* structures the thing it supports. The answers may vary. Its structuring effect may be great or small. For example, a mature tree may be supported by a post to keep it from bending beneath the wind. Without the support, the tree would nonetheless continue to grow, perhaps at a slant, and it would remain more or less the same tree. In the case of water, the *support* has far greater structuring power: without a container, water has no form and cannot be grasped, isolated or identified. Water takes the shape of its container.

A *support* for ideas, content and information can be said to operate in much the same way. Content (data, ideas, information, etc.) can be considered independent and self-sufficient to varying degrees, depending on one's point of view. The role of the *support* is to receive it and structure it. Content may be likened to the tree – the *support* only keeps it from growing

83   Donald A. Norman, *The Design of Everyday Things*, 1st Basic paperback, New York: Basic Books, 2002.
84   See Trésor de la Langue Française informatisé, on line: http://atilf.atilf.fr/.

warped – or on the contrary to water, which is completely shaped by the *support*. In either case, the conceptual structure is the same. On one side, there is the content; on the other, the *support* which structures it and which *is* the medium, in a sense. For example, a print-ed book can be considered the *support* for writing; the book is what holds the writing and partially structures it. The same text could be conveyed in another form, such as parchment or a monitor. The various *supports* are therefore understood as different ways of mediating between the writing and its reception. First came the writing, and then the document in the form in which we are reading it. The document changes form depending on the *support*. A different *support*, a different medium produces a different mediation. Clearly, the extent of the difference will depend on the structuring power we ascribe to the *support*. In writing like the tree – the difference between reading it in print or on a monitor is not huge, or more like water – the difference is pronounced since water can assume any shape.

This is precisely the idea McLuhan critiques in *Understanding Media*. The media theoretician showed that a medium cannot be separated from its content. The understanding of *support* and content outlined above raises two conundrums: what would writing without a *support* look like, and what would the *support* be if it bore no writing? There is no easy answer to this two-pronged question. We can readily imagine a bottle without water, but what about a book without writing or writing that is inscribed on nothing? This takes us to the heart of the debate about media that has followed from McLuhan's analysis. If a medium establishes a mediation between two things, it becomes difficult if not impossible to separately identify those two things.

Bruno Bachimont considers digital a *support*. He observes that things that exist in non-digital form (such as books, videocassettes, photos) can be transferred to the digital realm, calling this migration 'recontextualisation'.[85] The *support*, then, is the context in which the content appears. According to this logic, there is always a *support*. There is no writing or video without a *support*; they have only moved from one to another. But still, how is it possible to speak of transferring content from one *support* to another unless we are able to bracket the *support* and focus on the content? In other words, if we say we are transferring a video from cassette to digital, we assume the video exists apart from its holder. Or, we must assume that the holder itself has been transformed, which amounts to saying that the mediation is a mediation of a mediation, as McLuhan suggests when he says that the content of a medium is always another medium: the book contains the writing, which contains text, which contains language, and so on *ad infinitum*. In practical terms, it amounts to essentializing mediation, to making the mediation process itself an object, a substance, more than a *support*. Ultimately, this substance – whether it is a book, a screen or a cassette – supports only itself. Can we still call this substance a medium if the things it mediates no longer exist?

These considerations suggest that the very idea of *support* is problematic for intermedial discourse: Bolter and Grusin's concept of remediation already underscored this difficulty,

---

85   Bruno Bachimont, 'Le numérique comme support de la connaissance: entre matérialisation et interprétation' in *Ressources vives. Le travail documentaire des professeurs en mathématiques*, ed. by Ghislaine Gueudet et Luc Trouche, Paideia, PUR et INRP, 2010, pp. 75-90 http://hal.archives-ouvertes.fr/hal-00496590 [accessed 12 November 2012].

and McLuhan elucidated the impossibility of the unmediated object, for mediation is always a mediation of something. Therefore, the medium does not produce a mediation but rather a remediation. In line with this logic, a *support* can only hold another *support* and can never truly support content.

Interface

The concept of interface reproduces the same theoretical structure. The prefix 'inter-' also refers to an in-between, in this case to the interstice between user and machine, or between a graphical platform and a code. If the *support* is the mediation between the pure idea and the idea incarnate, the interface is the mediation between inside and outside, the transition zone.

But, as Alexander Galloway shows, the window metaphor often used to describe the interface cannot fully account for its role and function.[86] While a window frames a relationship between two spaces, it does not explain what happens in passing from one to the other. The interface is an in-between that effects a transition between two sites and is in this sense a mediation. Where the concept of *support* posits the existence of a thing that takes a different form than it initially had because of the in-between constituted by the *support*, the concept of interface offers the possibility of transforming content in an intermediate space in order to adapt it for a different space. The transition zone between inside and outside is the place where content is transformed to make it intelligible on the other side of the threshold. To bring inside and out-side face to face, there must be a link, like the lens in eyeglasses, that serves as the 'inter-face'.

A simple example is the interface that renders code usable by human beings. To take an everyday instance, a webpage is text marked up in html (hypertext markup language). An html tag is not intuitively interpreted by human beings: the code's 'inside' must be transformed by mediation to make it usable on the 'outside' by the viewer. To do so, a software application, in this case a browser, serves as the interface and transforms the code into the graphically organized page that users see on the screen. We therefore have two poles – the code and the page displayed – and an in-between – the interface – which mediates between the two and enables the passage from one to the other.

This example raises a series of problems as soon as we endeavour to analyze it more closely. First, the identification of the two poles is dubious. As we saw with the *support*, when we go looking for the 'original content' we quickly realize there is no content that isn't already a product of mediation. In the case of html, the code plainly results from interpretation of another code, which transforms a series of 0's and 1's into alphanumeric characters through the process called character encoding. The computer cannot generate text directly; it can do so only through a mediation – an interface – between text and numbers, based on a defined character set (such as UTF-8).

We therefore arrive at other content, the series of 1's and 0's. But this content is also the product of a mediation: the numbers must be converted into a series of electrical signals (on

---

86   Galloway, *The Interface Effect*, p. 39/40.

= 1, off = 0). And the electrical signals must in turn be interfaced with the circuitry, and so on *ad infinitum*. So we're back to the structure of mediation described by McLuhan, and the concept of interface loses its meaning once we realize that anything can be defined as an interface: any point may be read as a threshold, as a passage between one thing and another. That is precisely the structure we described when we looked at the concept of hypermediacy. So the question is not so much whether there is a passage or whether we are in it but rather what specific effects the mediation produces in a given situation. In other words, we must consider whether the mediation determines the content in question, and if so, how.

Apparatus

The concept of apparatus[87] is more helpful than the other two notions we have discussed in understanding how a mediation produces something. The central question relates to what occurs in the instant of mediation. How does the moment of mediation determine the meaning and structure of the thing that is mediated? For example, how does printing a book determine its content? Or how does publishing an article on the Web determine its meaning, reliability and substance?

Grusin uses the principle of 'premediation'[88] to analyze the force of predetermination in mediation. The structural organization of television illustrates the workings of premediation: the frame is set in advance; the time allocated to each specific type of content, to commercials and to the entire palimpsest is predefined. This framework has a decisive effect on everything we see on the screen.

'Affordance', particularly as Donald Norman defines it in his work on design, is a variety of predetermination that formally parallels premediation. Affordance is what a tool suggests by its shape and structure. Every technical object carries within itself a predetermination of the action it will set in motion: a chair invites the user to sit on it, a doorknob to turn it, a button to press it.

These principles of predetermination are inscribed in the notion of the apparatus, a complicated concept that demands clarification. To begin with Foucault's definition, an apparatus is 'a resolutely heterogeneous body composed of discourses, institutions, architectural arrangements, regulatory decisions, laws, administrative measures, scientific statements,

---

87   We use the English word 'apparatus' for the French word 'dispositif' following the translators of Giorgio Agamben, *Qu'est-ce qu'un dispositif?*, Paris: Payot & Rivages, 2007, translated: Giorgio Agamben, *'What Is an Apparatus?' And Other Essays*, Meridian, Crossing Aesthetics, Stanford, Calif: Stanford University Press, 2009. This word can be translated in many different ways: 'apparatus', 'device', 'machinery', 'construction' or 'deployment'. After the definition of this concept by Michel Foucault and others, *Dits et écrits, 1954-1988. II, II,* [Paris]: Gallimard, 2001, the term has been used by many scholars with differents meanings. See, for example: Arnaud Rykner, *Le tableau vivant et la scène du corps : vision, pulsion, dispositif*, Hyper Article en Ligne – Sciences de l'Homme et de la Société, 2013 https://isidore.science/document/10670/1.0jcfgt [accessed 16 February 2019], Agamben, *Qu'est-ce qu'un dispositif*, Bruno Bachimont, *Le sens de la technique: Le numérique et le calcul*, 2010.
88   R. Grusin, *Premediation: Affect and Mediality After 9/11*, 2010 edition, Basingstoke England; New York: Palgrave Macmillan, 2010.

philosophical propositions, morals, philanthropy – in short, both the said and the unsaid.'[89] This understanding of apparatus, which can be compared with the concept of institution developed and popularized by Pierre Bourdieu, is a central term for Foucault. Giorgio Agamben provides a three-point synthetic definition, based on Foucault:

a. An apparatus is a heterogeneous set that includes virtually > > anything, linguistic and nonlinguistic, under the same heading: > > discourses, institutions, buildings, laws, police measures > > philosophical propositions, and so on. The apparatus itself is > > the network that is established between these elements.

b. The apparatus always has a concrete strategic function and is always > > located in a power relation.

c. As such, it appears at the intersection of power relations and > > relations of knowledge.[90]

This definition recalls Éliséo Véron's definition of a medium, cited above.[91] The apparatus is the network that binds together this heterogeneous constellation and it performs an organizing function. It is a set of discursive and technical elements that determines actions and behaviours. The apparatus structures the realm of possibility. For example, a classroom is an apparatus. A heterogeneous array of factors – the architecture, the arrangement of tables and chairs, the school's authority, the teacher's role, the subject matter's place in the curriculum and in the collective imagination, etc. – shapes what transpires in the classroom, namely the course. Each apparatus has its specific affordance and premediation: a classroom designed as a lecture hall will prompt the teacher to deliver lectures and inhibit the functioning of any other type of activity, such as a workshop.

Bruno Bachimont stresses the determinative and even prescriptive power of the apparatus, which he describes as a 'physical and spatial organization capable of producing and determining a process. The essence of the apparatus is that its spatial configuration determines temporal behaviour.'[92] In other words, the apparatus transforms space into time. The combustion engine illustrates the process: it is a 'physical and spatial structure organized to produce a temporal cycle.'[93] The same applies to the classroom: the arrangement of the room and the set of spoken and unspoken rules that define its meaning and function produce and delimit the actions that can be performed within its walls. In Bachimont's terms, the classroom is an apparatus because it is a space that produces time and action of the course.

The apparatus, then, is what produces a mediation; it might be said that it is located between the intention and the completed action. For example, let us imagine an academic class; the

---

89   Foucault and others, *Dits et écrits, 1954-1988. II, II*, p. 299.
90   Agamben, *'What Is an Apparatus?'*, pp. 2-3.
91   Éliséo Véron, 'De l'image sémiologique aux discursivités, From the semiotic image to discursivity'
      Hermès, La Revue, 13, 2013, 45-64 http://www.cairn.info/resume.php?ID_ARTICLE=HERM_013_0045
      [accessed 30 April 2017].
92   Bachimont, *Le sens de la technique*, p. 42.
93   Bachimont, *Le sens de la technique*, p. 42.

apparatus 'course' is located between the course imagined by the teacher and the course she delivers is the classroom, which determines and structures her actions. Indeed, it is impossible to conceive of a course without the space in which it is conducted, for the space establishes the 'course format'. Clearly, the space is not created *ex nihilo*; as it is part of a larger complex – the institution, its protocols, its history, power relationships, expectations, and so forth – that endow it with its structuring power. It may therefore be concluded that the apparatus determines behaviours and actions.

Proceeding from the concepts of *support*, interface and apparatus, we can now attempt to tease out the meanings of the word mediation. We will distinguish among four types: formal, communicative, phenomenological and teleological.

In a purely formal sense, any process that links two things can be called a mediation. This is the type of mediation we find in the notion of the *exaiphnes*. The transition from point A to point B presupposes a linkage between them, and that link is a mediation. The advantage of this definition is that it posits mediation as a fundamental structure of our connection to the world. The relationship between stasis and motion, discrete and continuous, content and container can all be conceived in terms of mediation. The problem with this definition is its breadth; there is nothing that is not mediation. It is therefore too all-encompassing to be useful as an operational definition for a theory of media.

From the standpoint of communication theory, mediation may be defined in Claude Shannon's terms, as the process by which a message is transmitted from a sender to a receiver. This model is particularly well suited to analyzing the semiotics of media practices; that is, for investigating the production and circulation of meaning via different types of media. A book carries the meaning of its content and conveys it to readers. One limitation of the communication approach is that it does not lend itself to analysis of forms that do not seem to involve the transmission of meaning, and yet clearly qualify as media practices in our view.[94] For example, most Web activities serve instrumental ends more than communicative purposes and are not readily reducible to an exchange of messages. When we buy a book online, we are not having a conversation with someone, but rather manipulating objects, just as if we had made the purchase in a bookshop. To be sure, the online transaction, or the bookstore purchase for that matter, can be considered a form of communication, but that amounts to saying that everything is communication, much as everything is mediation in the formal sense discussed above. We find ourselves up against the same problem of generalization.

A third possibility is to ascribe a more phenomenological meaning to mediation and regard it as the condition of possibility for perception. This approach was already suggested by Aristotle in *On the Soul* and has been picked up more recently by intermedialists such as Lars Elleström. To summarize, mediation is the process that makes perceptible to the mind something that would not otherwise be so. This definition has the merit of being wide enough to embrace a series of media practices without necessarily leading us back to communication theory. The problem is its anthropocentric – or, at the very least, anthropo- and zoo-centric

---

94   Craig Dworkin, *No Medium*, Cambridge, Mass London: MIT Press, 2015.

– bias. The rise of information technology forces us to question this limitation, for the digital arena is occupied in large part by machine-to-machine interactions that have nothing to do with perception. Can those interactions be excluded from the realm of mediation? Another difficulty with this approach is the existence of practices – the magic show being the most salient example – in which the effect of the mediation is to make imperceptible what would otherwise have been perceptible.

Lastly, there is a teleological conception of mediation: mediation is what makes it possible to reach an end. It is the tool used to perform an action, the hammer that drives in the nail. This is the dominant idea behind our discussion of *support*, interface and apparatus: mediation is what enables and structures the action. This approach is compelling but comes with its own flaws: it emphasizes that mediation is a determinant of action, which is certainly true, but neglects the fact that the instant of action is subject to an element of indetermination as well. The concept of performativity offers us a theoretical handle on that indetermination.

Performativity in Intermediality

Postmediatic intermedial thinking dismisses essentialist arguments and foregrounds necessarily situational processes of convergence. From this point of view, there are no media; there are only dynamics – or, in Dworkin's terms, media are not things, but rather activities.[95] Reality is no longer apprehended as a series of immobile objects endowed with an essence, but as motion itself. There are no things, only movement; no essence, only action; no media, only convergences that produce mediations.

Clearly, this is not a recent opinion; as we have seen, it was first propounded by Heraclitus, the philosopher of flux and motion. Not only is this point of view not new but neither is it confined to philosophy. Nietzsche's critique of ontological thinking invites a number of arguments that have informed not only intermedial thinking, but a wide swath of contemporary research in the humanities, most notably in gender studies. For example, Judith Butler's questioning of traditional conceptions of sexual identity in *Gender Trouble* has some bearing on our own field inquiry:

> The challenge for rethinking gender categories outside of the metaphysics of substance will have to consider the relevance of Nietzsche's claim [...] that 'there is no "being" behind doing, effecting, becoming; the "doer" is merely a fiction added to the deed – the deed is everything'. [...] [W]e might state as a corollary: There is no gender identity behind the expressions of gender; that identity is performatively constituted by the very 'expressions' that are said to be its results.[96]

From this point of view, the poles between which the in-between supposedly lies are only a *post hoc* projection, constructed after the event. There is no white chair or blue chair but only a motion, a process. Our language needs to discretize the continuity of the process and

---

95   Dworkin, *No Medium*, p. 28.
96   Judith Butler, *Gender Trouble: Feminism and the Subversion of Identity*, New York: Routledge, 2006,
      p. 25.

hence describe it after the fact by postulating that there was a white chair and then a blue chair. The two chairs are freeze frames, abstract crystallizations of a continuous movement that resemble the sampling of the aforementioned sound waves. In reality, there is only the process. The instant in which the motion occurs is inhabited by a force, a tension that cannot be essentialized. It isn't a 'thing' or a space.

Clearly, it isn't easy to describe this motion using our language, with its heavy essentialist burden; our words are also essentializations. Plato's *exaiphnes* can be interpreted as an instant that eludes the discretized timeline along which essences are strung. To describe the same theoretical structure, Nietzsche used the term *das Tun*, Bergson *élan*, Derrida *différance*.

If we bring this perspective to bear on our analysis of technology and media, then the interface can no longer be understood as the point of passage between two things. The interface becomes the only reality: the two poles are produced retrospectively by the interface. Therefore, mediation cannot be conceived as a negotiation between two ontologically pre-existing poles, for those poles appear only belatedly, subsequent to the action which actually occurs in the interface.

As we know, the word 'media' is often used to essentialize mediation and frame it as a point of passage between two pre-existing poles. Rejecting the previous poles leads us to conclude, in agreement with postmediatic thinkers such as Galloway and Dworkin, that there is no such thing as media; there are only practices. In this case, there is no longer an in-between for there is nothing for it to be between. There is nothing but action transpiring in the continuum of reality. It is only in meeting the requirements of language that we crystallize actions by freezing and naming them retrospectively. The naming of media – radio, television, the Web, theater – is an example of this process. We shall return to this point in Chapter 4.

As we have argued, assigning priority to action and refusing the traditional concept of media probably constitutes the key shift in intermedial studies in recent years and marks the transition from the mediatic to the postmediatic phase. To sum up the logic of the postmediatic approach, we might paraphrase Judith Butler by replacing 'gender' with 'media' in the passage we have quoted: 'There is no medium behind the performance; media are performatively constituted by the very expressions that are said to be their results'.

The concept of 'performativity' – which focuses our attention on the action, on what is happening in the here and now, and links it to its conditions of possibility – proceeds from the same logic as the concept of media convergence that we defend here. The conditions from which it emerged, the tensions from which it sprang and which it has spawned, are akin to those that gave rise to both phases of intermedial thinking, and can similarly be traced back to the Platonic intuition of the *exaiphnes*.

Performativity as a Component of Action

While the concept of performativity originally emerged in philosophy of language, it bears relevance well beyond language, especially in the field of theater. Performance studies have contributed to some of the more promising theoretical developments of our day, as François

Cusset observes in *French Theory*.[97] Cusset argues that the American reception of the thinkers broadly referred to as 'structuralists' and 'poststructuralists' (notably Foucault, Derrida,
Deleuze and Barthes) sparked tremendous theoretical vitality in the United States while
stalling the debate in the French-speaking world. In the postscript to the book's second edition, Cusset notes that the arrival of performance studies in France at the end of the 2000s
ushered in a renewal of theoretical debate in the French-speaking world and is now driving
an international surge of new theoretical discussion. Performance studies can therefore be
regarded as one of the mainsprings of contemporary theoretical reflection.

The concepts of performance and performativity[98] are increasingly gaining traction across
fields of theoretical analysis: in theater, literature, philosophy and political science. However,
clear definitions that would enable us to distinguish between the two and pin down their
respective meaning sometimes appear to be lacking. We will attempt to elucidate such definitions here.

More specifically, we will look at what we have called 'action' and examine the relationship
between action and performativity. Every action is determined to a greater or lesser degree.
Depending on our analysis, some actions may appear wholly determined, others entirely free.
We have discussed the concepts of premediation and affordance, and considered how the
types of determination they imply influence media devices. We will now turn to the structure
of action to seek the non-determined element, the component of action that permits a margin
of freedom. We will call that element 'performativity', but we assign a more precise meaning to
the term than what we find in Austin or in performance studies. To understand how mediation
can be considered performative, we must first develop a theoretical analysis of performativity
as a component of action, and define the difference between performativity and performance.

In this discussion, it must be borne in mind that the determined and non-determined elements of action cannot be separated in reality. This is a discursive distinction that serves an
epistemological function, clarifying our understanding of the structure of action, and is not
rooted in any ontological divide.

Drama and Performance

We will begin our dissection of action by returning to the theater and demonstrating that action
on stage illustrates more generally the action's structure. Action in the theater presents us with
an evident paradox. On the one hand, we can think of it as the quintessential form of action:
overt, visible, public. At the same time, it appears fake and fictional, which is why the word
'theatrical' has taken on a negative connotation. What do actors do? According to one interpretation of dramatic action, an actor does everything but act, in the sense of performing an

---

97 François Cusset, *French Theory: How Foucault, Derrida, Deleuze, & Co. Transformed the Intellectual
    Life of the United States*, Minneapolis, Minn. [u.a.: Univ. of Minnesota Press, 2008.
98 For an overview of the two concepts and their history, see Nicholas Cotton, 'Du performatif à la
    performance' *Sens Public*, 2016 http://www.sens-public.org/article1216.html [accessed 20 February
    2018].

action. This is a striking contradiction: the word actor, which specifically denotes the subject's relationship to action and doing, also refers to an essentially passive function. Actors do not act: they don't love, or kill, or die on the stage. Actors don't do anything because everything has already been determined: the action on the stage is not an original but a reproduction, a sham. It is a feigned and deceptive 'liveness'. Theatricality, then, is a fiction, a form of mimesis; it designates representation as opposed to presentation.

Another aspect of traditional drama is the degree to which the actors' actions are determined. Everything is often decided in advance. Drama consists in recreating, on the stage, something that has or could have already occurred. The action happened in the past and is only evoked, recalled, imitated in the theater. The script crystallizes and standardizes the action. A flawless drama, then, is one in which nothing happens: it is a perfect reproduction, a replica that is indistinguishable from the original and, precisely for that reason, entirely false.

Performance lies at the other end of the spectrum: there are no actors in a performance precisely because there is action. The performer isn't reproducing anything: she acts, she drinks, she eats, she loves, she moves and she suffers.

But this opposition, which underpinned the abandonment of theatricality in the theater in the 1980s and 1990s in favor of performativity, cannot withstand scrutiny. The question of the determination or indetermination of action is more complex and irreducible to any binary pair. The true polarity – though strictly speaking it is no such thing, at least not in the conventional sense of polarity – must be sought elsewhere, in the opposition between the determination and performativity of action. Those two elements, which coexist to vary-ing degrees in any action, define what happens on the stage as drama or performance. All theater is partially determined, from the most faithful rendition of a script to the most improvised performance.

What is true on the stage applies to all action. The dramatic model calls for particularly stan-dardized action, while the performative model leaves more room for indetermination. But in every case, on or off the stage, our actions contain an element of determination, which we shall now consider in greater detail. That element of determination has multiple and varied aspects, but we will confine ourselves to three that we believe to be of particular importance: the spatial frame, the symbolic context and the physical constraints. These factors all fit into the framework that Grusin calls premediation.

When a stage is used for a play, a playground for a baseball game, a classroom for a course or a bedroom for a nap, the spatial and architectural environment limits and structures the action. The architecture, understood in its broadest sense as the organization of space, func-tions as an apparatus; it predetermines – and premediates – the spectrum of possibilities. A chair predetermines the act of sitting down, a door the act of opening and closing, and so forth. 'Natural' spaces, necessarily perceived through a cultural lens, are also endowed with this power of predetermination, lending them an architectural dimension. A lake beckons us to swim in summer or skate in winter.

The second determinant is the symbolic context of the action. In the case of the theater, it is what was written, put in place or more generally planned prior to the drama or the performance, *i.e.* the script. Outside the theater, the symbolic context may also be a 'script' that structures and determines actions. That script consists of all the texts that make up the social and cultural framework: mores, traditions, the social apparatus.

Lastly, there is a series of mundane logical and physical necessities: physiological requirements, physical forces and limitations, logical principles (it cannot be raining and not raining at the same time).

These three factors, to which we shall return in the next chapter, determine and circumscribe the action. From a strongly deterministic point of view – such as the one implicit in the concept of apparatus as defined by Bachimont – no non-determined space remains: if we knew all the constraints, we would be able to predict every action. Spinoza's *Ethics* provides a well-known example of this claim; for Spinoza, an action is the result of all the constraints by which it is produced. There is therefore a strict causal relationship between the preceding situation and the ensuing action.

The Temporal Structure of Action

Determinism can take various forms: physical (such as Laplacian determinism, which holds that if we knew all the physical variables in play at time T, we could predict the state of the world at time T1),[99] logical (Diodorus Cronus's master argument, according to which, based simply on the principle of non-contradiction, we can establish which propositions are true and which are false at any given moment, and therefore into the future),[100] or fatalistic (there is a predetermined ending, like the script of a play).

All these varieties of determinism are rooted in a specific interpretation of the time sequence in which the action is embedded. They assume linear time, consisting in a series of discrete units from which particular moments, linked by the principle of causality – a string of cause-effect relationships – can be picked out. In other words, the deterministic conception of action is entrenched in an essentialist understanding of reality: there is a 'before' – the script – and an 'after' – the playing out of the script. We therefore return to the two poles discussed earlier, and the essentialist perspective postulating that they are ontological precedents to the in-between that connects them.

Ontologically, then, the essence of the 'before' – the script – and the 'after' – the performance – must already be given, prior to any action. To stage a drama, there must already be a script and an idea of what it would and should look like when performed on the stage. From an

---

99 Robert C. Bishop, 'Determinism and Indeterminism' *arXiv:Physics/0506108*, 2005 http://arxiv.org/abs/physics/0506108 [accessed 28 May 2018].
100 Aristotle outlines the argument in *On Interpretation* 18b35. For commentary, see Hervé Barreau, 'Le Maître argument de Diodore: Son interprétation traditionnelle, sa signification historique, sa reconstitution contemporaine', *Cahiers Fundamenta Scientiae* 46, 1975, pp. 1-51; and G. Anscombe and M. Elizabeth, 'Aristotle and the Sea Battle: De Interpretatione, Chapter IX', *Mind* 65, 1956, pp. 1-15.

ontological point of view, the two poles have priority over the action on the stage that runs counter to our conception of the chronology of action.

Approaching the question from a chronological angle, Paul Ricoeur describes a sequence of mimetic action consisting of an original, a mediation, and then a copy. There is an initial pole (the original) and it is from the mediation – the in-between – that the second (the 'copy', or the representation) springs. But while there is a shift in perspective between the ontological and chronological points of view, they spring from the same essentialist conception of the order of things. Both rest on the discretization of reality and agree on the primacy of objects.

These two foundation blocks are the nub of the problem. By positing the existence of two poles, the essentialist approach forces us to attach the action to one of them, specifically the second, the one that comes chronologically after the first. The action is understood as the simple effect of a determining cause. Applied to the theater, this logic casts the performance – determined by that which preceded it and is its cause, namely the script – as the action.

But how can we conceive of the script as a separate, pre-existing entity? Contrary to Barthes, recent work in dramaturgy by Joseph Danan convincingly shows that the script also fully participates in theatricality; it already bears within itself the conditions of its performance. In a way, it is a product of the performance because, to enlist Grusin's concept once again, it is 'premediated'.

If removing the script from the creative dynamic to which it belongs served only to reify a gesture, the consequences would be relatively limited. But the attempt to fragment a flow, to oppose action and object, to separate the inseparable, which is characteristic of the essen-tialist approach, has more dire effects. Those consequences urge a non-essentialist position which holds that script and performance can be discussed only from the standpoint of action: there is no object outside the action, and for there to be an object, there must be an action. For there to be a script, there must be the action of writing, or reading, or playing.

Returning to the example of the chair, there aren't two chairs and something that connects them. There is only an action, the act of painting a chair blue. It is the only reality. The two chairs, white and blue, are discursive constructs. They are a response to the action of paint-ing, a *post hoc* mental configuration. Similarly, it could be said that in the theater the action is not the performance of a script but what happens on the stage, from which script and performance emerge to form a dynamic, indivisible whole.

Clearly, the non-essentialist approach poses a major challenge, for the real, continuous instant in which the action occurs is ungraspable and undefinable. It remains mysterious and escapes all the varieties of determinism implied by essentialist thinking. It is in this instant – in the *exaiphnes* – that the performativity of each action is to be found.

We propose to further develop this perspective and frame a different, non-essentialist, non-de-terministic conception of action and hence of mediation. To do so, we must first define the difference between 'performance' and 'performativity', which are often confused. Performance is an artistic rendering while performativity is a property of action. Any action is, to some degree, performative.

The Undecided

Performativity is the element of each action that eludes determination. It is what happens in the instant of action that could not have been foreseen. Let's return to the example of traditional drama, which would appear to be the least performative variety of action, since everything seems to have been planned in advance (barring an element of improvisation). The actors recite the text without changing a word. They have rehearsed it many times to find the best delivery, and just the right intonation, for each line. Their movements were blocked beforehand. And yet, in the moment when they are on stage, acting, something happens that transcends the predictable. A singular tension is created precisely because what happens on the stage is happening *there*, and no amount of rehearsal can strip that moment of a margin of unpredictability. There is always a gap between what was determined and what occurs in the *exaiphnes* of the action. It is in this sense that an actor acts. All was decided and yet an undecided slice remains. That space is precisely the locus of performativity, for the undecided element is what remains to be determined, what remains open, unpredictable.

There are therefore two distinct but complementary ways to view an action. From an essentialist position that conceives of reality as discrete, we may focus on what predetermines an action, or we can try to understand it as part of the flow of reality, as that which fills the *exaiphnes* and always remains undecided.

Clearly, both positions are tenable. It cannot be denied that every action is determined and, at the same time, retains a margin of indecision. But viewing action from the standpoint of performativity is, first and foremost, an ethical and political imperative. Those twin demands prompt us to redefine our terms. From a theoretical point of view, we may choose the approach that suits us best and focus on analyzing the determinants of action to understand its causes, if choose to. From a political point of view, we need to seek the performative element of action because that is where we will find the zone of freedom and responsibility. We cannot be responsible for something that has already been determined, but only for what is as yet undecided, where our actions can sway the outcome. Responsibility prevails in the *exaiphnes*: only in the non-determined space where I enjoy a margin of freedom can I be accountable.

From the ethical point of view, the key question is: precisely what remains undecided? The ethical approach probes the performative aspect of an action to delineate the zone of freedom in which responsibility is possible. If we are not responsible for what has already happened, for what is finished and immutable, but only for what can still be done differently, then performativity is the condition of possibility for ethical choice.

Viewed through a political lens, all norms and apparatus of power are determinants of action. But an action can have no political significance if it is entirely controlled by those norms and apparatus: it assumes political meaning when it seeks the empty space, the zone of indecision that allows for creative normativity. Performativity disturbs norms because it overflows from their determinative power and produces something new, unexpected and therefore, to some degree, subversive.

These ethical and political considerations are the rationale behind the narrower sense we assign to the term performativity. First, it is important to detach performativity from performance as an artistic practice, although the two are historically linked, and to distinguish between the two meanings of 'performative', wherein one is bound up with performativity and the other with performance. In common parlance, performativity and performative relate to the general idea of action in progress, of that which is being done. Indeed, 'performativity' and 'action' are often used interchangeably. This very broad usage is problematic. While every action has a performative dimension, the performative is not coextensive with action. In our definition, performativity denotes only the undecided component of action, which is also its normative component, in the sense that instead of perpetuating existing norms, it produces new ones.

By thus distinguishing action from performativity, we can avoid the debate about the determination and indetermination of performativity that Judith Butler sums up in *Excitable Speech*.[101] She outlines the limits of the approach that originated with J.L. Austin's *How To Do Things With Words*,[102] a seminal text in the development of performance studies. For Austin, utterances are performative when they 'do' an action that has been predetermined by the situation in which they occur. The sentence 'I pronounce you husband and wife' is performative because it creates a marriage, in accordance with the pre-existing codes and conventions that determine what a marriage is and how it is contracted.

Butler argues that the performative can, on the contrary, produce something novel, something that transcends and challenges the situation in which it occurs. But this does not mean the situation imposes no limits. For Butler, there is an interplay between situational determination and norm-creation in the performative.

We agree that action has a dual nature, but we will use the term 'performative' to refer only to its non-determined side. As we have argued, the line between the two components can be drawn only discursively, for the performative and the determined cannot be sundered in reality. Nevertheless, we must consider them separately if, for ethical or political reasons, we wish to examine the undecided element of action.

Mediation as Performativity

What does this perspective mean for mediation? We have seen the essentialist and non-essentialist interpretations of the term. From a non-essentialist point of view, mediation is a way to understand the *exaiphnes* as an instant that belongs to neither the before, nor the

---

101 Judith Butler, *Excitable Speech: A Politics of the Performative*, New York: Routledge, 1997.
102 John Langshaw Austin, *How to Do Things with Words: [The William James Lectures Delivered at Harvard University in 1955]*, ed. by James Opie Urmson, Cambridge, Mass: Harvard Univ. Press, 2009.

after. Mediation is what happens in the moment: it is the creative instant of action. In this sense, it is pure performativity.

From an essentialist point of view, mediation can be described as something between two moments, two objects, and so forth. There is a moment A and a moment B; and mediation is the path from one to the other. Thus, as we have discussed, mediation is itself essentialized and reduced to a determined, pre-established apparatus. In the process, the indecision that inhabits the moment of mediation is lost.

Let us take an interface as an example. It can be described in two opposing ways. From the determinist / essentialist point of view, it is what lies between code and user. The code and the user were already there, determined and well-defined. The interface links them but is wholly determined by the nature of the two things it connects. The code is already written, the user is already in front of the screen with expectations, abilities and needs. From the characteristics of code and user, we can deduce the interface. Its function is pre-established: what happens in the mediation reproduces what was already plotted. Similarly, the essentialization of the interface compels us to conceive the action that occurs within it as entirely determined – by its premediation, by its affordance, by the fact that it is an apparatus, in Bachimont's sense.

If our interest is in the undecided element, we must frame the picture differently. Mediation becomes the starting point of the analysis, for it is all we have. In the *exaiphnes*, there is neither user nor code, but only the interface. The user and the code are produced by the interface: they make their appearance only because there is an interface that connects them. All that seemed to be decided and determined is in fact open: the mediation is located in the undecided zone and is therefore performative. Or, more precisely, the mediation contains premediated and undecided aspects that are ontologically inseparable, but we must consider them separately in order to distinguish the element of freedom in every action.

This approach is particularly pertinent to the analysis of digital practices, as shall be seen in the final chapter. Much of the discussion around information technologies has centered on their normative dimension, how algorithms and technical devices premediate action to the point of transforming users into automatons.[103] But while focusing on the predetermined aspect of technological practices illuminates a basic feature of their operation, it is politically perilous, as it forecloses responsibility and gives rise to a discourse imbued with paternalism. It is true that any use of a platform is premediated by its structure, but there are also emancipatory possibilities insofar as our behaviours can etch new norms for that platform. We are conditioned, to be sure, but we also condition at the same time. We may be exploited, but that exploitation bears within itself the potential for normative agency.

In the following chapter, we explore this idea and illustrate the relationship between mediation and performativity, which we will attempt to elucidate using the concept of mediating conjunctures.

---

103 Matteo Treleani, 'Le spectre et l'automate. Deux figures du spectateur' in *D'un écran à l'autre. Les mutations du spectateur*, Delavaud, G. Chateauvert, J., Harmattan, 2016.

MEDIATING CONJUNCTURES

Between Multiplicity and Identification

The idea of mediating conjunctures began with an observation: the terms used in the field of media studies are problematic because either they relate to an essentialized vision of mediation phenomena, or else they are vague in that they have no operational value. In the previous chapter, we defined the difficulties created by notions of *support*, interface and device. And yet the most problematic notion is the term 'medium' itself.

We have shown that this term is tied to the intermedial phase that we have named 'mediatic'. This notion, as well as its associated terms, falls under an essentialist vision of mediation phenomena. This is why, based on the postmediatic phase, researchers such as Galloway propose that we simply abandon the concept of medium and concentrate on mediation. We are likewise convinced of this necessity, but we believe that we should also specify what the mediation process refers to.

One of the founding hypotheses of this book is to think that there is no initial element that will thereafter be 'mediated': there is nothing that comes before mediation. Bolter and Grusin apply the same principle: there is nothing prior to the act of mediation. This initial aspect of mediation implies that we cannot think the real as a layering of mediations.

But this approach has a limit, as it provides a definition of mediation that is simultaneously too vague and too particular. Too vague because it ignores both the production methods relating to the mediating operation and the moment that it is produced; too precise because it presupposes that there would only be one operation, implying that it is, generally, a singular process: we refer to mediation and not mediations. And yet, if we were to concentrate on what is produced at the time of each action, what occurs appears a little more complex. There is no mediator; there is instead an ensemble of elements in play at the time of the action and these elements (forces, human and non-human agents, contextual aspects...) intersect, merge, oppose, meet and double in a continual movement. It is this desire to move away from mediation as an objectifiable and singular process that leads us to resist the idea of 'medium' that refers to prefabricated mediation forms that are already structured and identifiable.

In the instant of the action, if we observe the interplay of these elements, we can identify particular conjunctures. These conjunctures are moving combinations that form the elements in play at the time of the action. This is what we call 'mediating conjunctures'.

The notion of mediating conjunctures allows us to find a median path between essentialism and non-essentialism, and therefore to not reduce mediatic practices to defined and pre-defined units, but to otherwise consider them as moving dynamics, without abandoning the possibility of identifying the specific characteristics of these practices. In completing Galloway's definition by which there is no media, but only mediation, we can say that, in reality, there is no mediation that is singular and identifiable, but mediating conjunctures.

Mediating conjunctures are necessarily plural. They are the moving contexts – and thereby irreducible to *one* context – of action, the dynamic configurations that an ensemble of interplaying forces and agents in a determined space adopt at the moment of action. These mediating conjunctures are what make the action possible, but they are at the same time what produce it. This paradox warrants clarification: on the one hand, mediating conjunctures are the environment where the action takes place, they are what in part condition and shape it; on the other hand, they can only be formed because the mediating dynamic has already been set off.

The concept of stigmergy, that we borrow from Ollivier Dyens,[104] appears particularly useful for explaining this singular logic. Originating in the field of biology, stigmergy describes a form of self-organization. It is based on a simple principle: that a trace left by an action in a given environment stimulates the completion of a subsequent action in the same environment, regardless of whether this new action is accomplished by the same agent or a different agent. The movement of schooling fish offers a good example of stigmergy. The school's shape is the outcome of all of the fish's movement but, at the same time, it is also the cause: each fish's movement is determined by the shape of the school. In other terms, the notion of stigmergy allows us to think about the premediated aspect of each action and its performative dimension. If we could take into account the ensemble of perspectives from which the school of fish is observable, we would have an exhaustive account of mediating conjunctures present in one precise moment. This is, however, practically impossible and we must, as often as possible, concentrate on the most precise elements. For example, we could disregard everything surrounding the schooling fish and limit ourselves to trying to understand the mediating conjunctures that are related to its internal dynamics. But we could also widen our point of view and identify another mediating conjuncture that would include, for example, a shark in the vicinity of the school, whose action would obviously influence the shape of the school and the behaviour of the fish. We could also observe the ensemble of dynamics in play in the sea or land area, or even concentrate on the simple interaction between the fish involved in the school. It is clear that these possibilities are infinite and that each is legitimate: it is therefore the plurality of these points of view that determines the plurality of mediating conjunctures. One particular mediating conjuncture is a discursive abstraction, made as a result of one particular perspective.

The notion of mediating conjunctures allows us to avoid the pitfall of conceiving of mediation as an exclusively homogenous device determining, structuring and producing the action. In other terms, the idea of mediating conjunctures allows us to resolve criticisms regarding notions of *support*, interface and device, as covered in the second chapter. We develop the concept of mediating conjunctures because the word 'mediation' suggests the existence of a single mediating element, structured and defined that – as the combustion engine described by Bachimont – would transform space into time, completely determining every event produced within the frame of the mediation. As in the case described by Matteo Treleani,[105] who

---

104 Ollivier Dyens, *Enfanter l'inhumain: le refus du vivant*, Montréal: Triptyque, 2012.
105 Matteo Treleani, 'Le spectre et l'automate. Deux figures du spectateur', in *D'un écran à l'autre. Les mutations du spectateur*, Delavaud, G. Chateauvert, J., Harmattan, 2016.

asserts that a user before a web document is as an automaton, that is to say that in a nearly mechanic way he performs actions that are measured by the mediating device: one click on a link, then another, and one sees a series of actions framed by a mechanic and algorithmic procedure. The idea of mediating conjunctures is, on the contrary, based on the conviction that what mediates does not exist before mediation. There is no prefabricated tool which, all of a sudden, mediates the actions by shaping them: the mediating contexts arise at the same moment as the action, and they are multiple and moving, and this is why we can understand them as operating at the conjunctures of many elements.

Thus understood in their multiplicity, mediating conjunctures cannot be associated with an essence. Trying to grasp and characterize them in a stable and fixed manner can only occur at the risk of disregarding the movement by discretizing its continuity. Therefore, we can select a determined time interval – a second, a minute, an hour, but also a period – and try to identify the characteristics of a mediating conjuncture during this time interval. Alternatively, as in the schooling fish example, we can turn our gaze to a determined part of the space. But this part will never be motionless, as it is our gaze that disregards the changes by behaving as though certain characteristics of the mediating conjunctures remain more or less the same during this time interval.

We will return to the example of theatrical action to which we have already alluded. When we refer to 'theatrical' action, we use the adjective in order to refer to a particular mediation that shapes and contextualizes the action. The epithet 'theatrical' refers to a mediating conjuncture that has been identified, grasped and described. This is only possible because we have limited our gaze to one predetermined part of time and space: we have disregarded other shifting mediating conjunctures in order to concentrate on one point of view and to isolate an ensemble of factors, the forces and the agents that we gather under the term 'theatrical'. This mediating conjuncture ends up crystallizing and making way for a particular medium: theater. The notion of mediating conjunctures helps us to understand that this is an essentialization that thereby consists in naming this activity 'theatrical' according to the needs of language. But this does not correspond to the real state of things. In other terms, the adjective 'theatrical' is a word that serves to identify a complex and shifting combination of elements that form the particular context for an action, at the same moment that it is produced. This context cannot be a closed and stable device that remains unchanged for all 'theatrical' actions, be it present actions or passed actions.

In this sense, the difference between a theatrical action and a film action or a reading action, or even a love action, is only a difference of degree and not of nature. It is, in other words, a difference that can be situated on a continuous spectrum: the different mediating conjunctures are not divided and separated in a discrete manner. For example, a theatrical action in the context of a contemporary play is closer to a dance performance of the same period than a Molière play. This perspective simultaneously allows us to take into account the premediation that exists within each media practice without essentializing it. This is at the heart of the notion of mediating conjunctures.[106]

---

106 The notion of mediating conjunctures, as we conceive of it, could be likened to that of the 'cultural

Let us return to the example of theatrical action in order to explain this point of view. If we imagine an actress, she is on stage, playing the role of Phaedra. She speaks, she crosses the space, she looks, she moves her hands, her body. There is something in the process of being produced in the same instant that the actress acts: it is an action. When the actress is in the process of doing something, her action is what counts the most, the most important thing. The fact that this action takes place in a theater, on a stage, is one of the aspects of this action, but it is not the only one and, above all, it cannot be separated from all the other aspects involved in the action. And yet we have said that it is a 'theatrical' action, as this action is characterized by a context, an art and a series of forces in play at the moment that it is produced. There is a stage, a space in which the action must take place. There is then a theater: a spatial device that includes the stage and allows an audience to observe it. There is also a text – Phaedra's replicas – that the actress has learnt and is in the process of performing. There are a series of rehearsals during which the actress has practiced her role and perfected it. There is also a certain light, a certain temperature, the actress's physical condition, a particular audience, a sound at the end of the room, a particular and changing 'atmosphere'. As an ensemble, these elements cross one another in creating particular conjunctures. To identify some of these elements in play is evidently an abstraction: it is impossible to demarcate in a predictable manner the perimeters of observation. For example: we can take into consideration the health conditions of the actress and tie them to a history of her illnesses since childhood – or place them in a genealogical lineage; we could do the same thing for the spectators; we could take into account the weather conditions and situate them in an ensemble of atmospheric phenomena... and yet each action, theatrical or not, is produced within the context of a particular ensemble of mediating conjunctures. In the case of a 'theatrical' action, we find a certain number of recurring elements that are more present than others: the stage, the theater, the audience, the text, the rehearsals... These interrelated elements contextualize the particular type of action that we consider theatrical. From there we could isolate these elements and say that they are a form of mediation or, even further, that they constitute one defined media: the theater. The theater would be a medium as it corresponds to a more or less stable ensemble of – seeing as this varies – mediating conjunctures. Obviously, this formulation is justified and can be very useful. When we buy a ticket to go to the theater, or when we talk about the play *Phaedra* that we saw the evening prior, the reality of being able to say that it was in fact the theater is crucial. But this gesture involves isolating a certain number of elements that constitute a mediating conjuncture, and naming it is always an arbitrary gesture. Let us say, for example, that the actress in question performs on the street, doing without the theater building and the traditional stage, and a group of people stop to listen to her. Could we say that this is not theater? And if we say that

series' developed by Gaudreault and Marion., André Gaudreault and Philippe Marion, *The End of Cinema? A Medium in Crisis in the Digital Age*, New York: Columbia University Press, 2015. What we are trying to highlight with our expression is the dispersed aspect of mediating conjunctures: a multiplicity of mediation elements that cannot – or not necessarily – be classified and organized into a single 'series'. To identify the development trajectories that would be the most important, or more noteworthy than others, in fact risks returning to a form of essentialism. The utility of the cultural series is that it finds a relationship between different media experiences that can be integrated into the same history. In the case of mediating conjunctures, it is not a matter of accepting that every history of each media practice is only one of many avenues of possible interpretations.

she does not speak in French, but in Italian, thus distancing herself from Racine's text, or even that she improvises a text that speaks of Phaedra but has nothing to with the known texts, would we say that this is not theater?

We will later return to these questions, but we outline, for now, that mediating conjunctures are never stable, and they never occur in exactly the same way; they are always moving, always changing. This does not prevent us from trying, for sake of analysis, to recognize elements in the movement of mediating conjunctures in order to bring together a certain number of actions, by characterizing them as being part of the same type. The fact that mediating conjunctures are not graspable in discrete units also depends on another factor that which we have just described with the help of the stigmergy concept. The action that is produced in the mediating conjunctures is also a part of mediating conjunctures, as it is one of the elements in play during their constitution. The actress that plays Phaedra participates, by her action, in the mediating conjunctures within which she is placed. This is precisely what the notion of stigmergy implies: the movement and the behaviour of each animal contribute to the production of a dynamic spatial structure – the school of fish or the ant hive or the spider's web, but this spatial structure itself conditions the behaviour of the animals. There is no 'school' without the movement of each fish, and yet the movement of a fish or of a particular bird is initiated by the structure of the school. It is obviously the movement of the spider that determines the structure of the web, but at the same time the webs architecture determines the movement. The two occur together because one forms part of the other. The action that is produced in the mediating conjuncture is at the same time conditioned by its mediating conjuncture – as the latter is its context – but, in turn, it conditions and shapes the mediating conjuncture. Without the actress who acts, the mediating conjunctures would be different.

Another example could help to understand this complex relationship between mediating conjunctures and action. We are, at this very moment, in the process of writing. Our writing action is produced at the crossroads of a series of forces in play at this very moment. There is a room with a particular lighting. There is a computer – an entire series of organized techniques and technologies, a tradition of writing, a play with letters, a keyboard, a word processor with its complex history and its relationships with other older word processors, a typewriter, handwriting. There is our ability to write: the fact that we have learnt a writing technique – first by hand, then we were able to adapt this technique to a machine and eventually to a computer, to different word processors, to different keyboards, to different writing situations and finally to this particular writing situation that is occurring at this very moment. There is a language, English. There is a body of work that had been done prior to writing and that gave us ideas. There is the desire to write, at this very moment, and the possibility of doing so. All these elements are in place and they are the forces currently in effect, in this precise instant when we are in the process of writing. The history of writing or the history of our work to prepare this text has meaning only because they have produced a force that is in action at this very moment. And yet the fact that we are in the process of writing is, itself, a force in play. The fact that we write forms part of the mediating conjunctures that characterize this very instant. The proof lies in the fact that what we write, the way we write, our movements on the keyboard, but also the letters that we choose, the words that form, all condition what follows.

In other terms, our writing action is part of a context in which this writing action is produced. Our writing action produces the context that thereby renders it possible and shapes it.

The Dimensions of Mediating Conjunctures

Let us now try to analyze mediating conjunctures to provide an interpretation of their various elements. We can analyze these elements on the basis of three coexisting and complementary dimensions: a technical dimension, a discursive dimension and a practical dimension. The merging of these three dimensions is what we have described using the stigmergy dynamic. The ensemble of these dimensions characterizes all the moving forces that come into play in mediating conjunctures.

1. Each action falls within a particular technical context. Therefore, the actress who acts forms part of a heterogenous ensemble made up of techniques ranging from architectural approaches to the theater's construction, to sound devices, to acting, etc. A user who creates a Facebook profile acts within a technical context characterized by information protocols, machines, algorithms, ergonomics... As outlined, it is impossible to demarcate this context in a single way. There is a not *one* theatrical technique, given that the ensemble of technical approaches in play in each context is very heterogenous. Therefore, we refer to technology in order to make reference to this heterogenous assemblage that refers to the competencies and tools that, given their complex history, can hardly be considered a homogenous whole. We can say that this heterogenous ensemble of techniques creates a space in which the action then takes place, and we can thereby tie this technical dimension of mediating conjunctures to the spatial aspect determining the actions analyzed in the previous chapter. We will return to the examples of theatrical actions and writing actions in order to better understand what we call the technical dimension of mediating conjunctures. In the moment that she acts, the actress finds herself in a particular space. This space is the ensemble of relationships tying the stage to the audience – the orientation of seats, the height of the stage in relation to the auditorium, the placement of the theater wings –, but that also tie the different objects on stage to one another – the scenography, the position of other actors –, and that tie together the ensemble of the 'theater' structure with its external factors – the theater's location, for example. The actress's action, her acting, only has meaning because it takes place in a particular spatial context: she is on the stage, there is an audience watching her, she finds herself in a theater, this theater is in the center of, say, Montreal, or even in the suburbs of New York. And yet this space is not simply the ensemble of elements and forces present but, moreover, as we have said, the ensemble of relations that bind them together. This is why this space can be considered the technical dimension of mediating conjunctures; elements such as lighting or acoustics are part of the spatial context. The relationship between the actors and the audience is not defined simply by the distance or geometrical position of the stage in relation to the auditorium and the theater boxes, but also by the fact that a particular lighting device enables the stage to be seen by the spectators, and the sounds produced to be heard. All of the technical and technological devices in place form part of the spatial context

that produces them. If there is a microphone, or a video recorder, or a computer that connects another theater via videoconference, these apparatus also form part of the spatial context because they contribute to building relationships between objects.

2.   The technical dimension is necessarily accompanied by a discursive – or cultur-
     al – dimension, without which the other would not make sense. This is what we
     have demonstrated in our discussion on sociomedialities. It is not, as we henceforth
     highlight, an abstract or immaterial aspect of mediating conjunctures. The discursive
     dimension is made up of the ensemble of documents, texts, or other forms of traces
     on which our interpretation of the world is based.[107] These diverse documents, tradi-
     tions and traces give way to a series of interpretative tendencies that shape the action
     at any given moment. The actress acts because there is a text -- in our example
     Racine's text. She also acts because there is a tradition wherein the theater means
     something for a group of people or a society. There are texts that have been studied
     and transmitted, a learning that occurs at school – or elsewhere –, a familiarity with
     the device that is possible thanks only to the sharing of communal textual or informa-
     tional references. The ensemble of these references is shared by the audience –
     each member of the public in their own way –, by the actors, by the director. And
     these references are also shifting, as they do not cease to change and produce the
     context for the action. If we study these forces, we can identify particular tendencies
     that characterize a given moment, or a period, or a society. We could, for example,
     argue that a period is characterized by the strong importance of the concept of rep-
     resentation. Or again a certain culture is founded on a particular idea of the world's
     hierarchy. To identify this is, of course, a discretization, as these forces are in move-
     ment. Identifying a symbolic tendency is taking into account a norm at the center of
     a dynamic and changing landscape. The coincidence of a series of symbolic tenden-
     cies gives way to conjunctures that can be identified as given forms of mediation: for
     example, the theater. But in this instance, also, identifying this is an abstraction, a
     necessity of language – as we will see in more detail in the next chapter. Discursive
     aspects are fundamental in understanding mediating conjunctures also because they
     are not necessarily in harmony with technical features. Industry discourse relating
     to high-fidelity sound and image reproduction technology is an example. To under-
     stand the specificity of digital sound we must, on the one hand, understand technical
     sampling and discretization of analogue sound and, on the other hand, understand
     industrial discourse that – as opposed to technical realities – presents this sampling
     as a faithful representation of the continued realities of sound. CDs are, at the same
     time, a reproducing technique that reduces 'fidelity' to the reality of sound, but also
     a cultural object that, according to this industry's discourse, results in higher fidelity.
     This paradox is unsurprising given the complex and shifting multiplicity of mediating
     conjunctures. It is precisely what Éric Méchoulan proposes when he defines inter-
     mediality as a hermeneutics of *supports*: 'with intermediality, we are not interested in
     the hidden meanings beyond the text, but in the operations that unite matter, forms,

---

107 Regarding this aspect, see Maurizio Ferraris, *Documentality: Why It Is Necessary to Leave Traces*, trans.
     by Richard Davies, New York: Fordham University Press, 2012.

social usage, scholarly practices, the historicity of phenomena and authorizing institutions'.[108] In this sense, the *support* is no longer a stable premediating technique, but
a dynamic produced by the interlacing forces put in play in a particular moment.

3. Lastly, we must underline the practical dimension of mediating conjunctures, which
    refers to the fact that actions taking place form part of these conjunctures and struc-
    ture them. The action is made possible by the context wherein it is produced but,
    at the same time, it is precisely what puts it in place. If the actress's role had not
    existed, there would not have been, in this precise case, a play, there would not have
    been a stage, there would not even have been lighting, or a microphone. And, more
    generally, without acting, there would have been neither a theater nor theatrical de-
    vices. In the same way, if there had been no users that created their Facebook profile,
    there would be no Facebook. We will later return to this fundamental aspect.

Let us illustrate the coexistence of these three dimensions with the help of an example,
that of digital photography. We could argue that digital photography is a separate *medium*
that emerges as a result of the development of preceding *media*, while obviously remaining
separate from these phenomena. Indeed, it is a particular technique that puts in place specific conditions that can thereby be described by a series of essential and defining features.
But separating digital photography from preceding *media* is not commensurate with actual
practices. We could say that digital techniques allow us to immediately capture a particular
moment, without the need of posing subjects, in at once obtaining the photographic result
of the photo – which does not need to be printed in a laboratory. But we could also consider
things differently by trying to capture the mediating conjunctures that constitute the photographic moment (the moment wherein one triggers the shutter open). We will see that a series
of ideas, of tendencies, of conditions constitute the elements put into play which trigger the
shutter open. These tendencies are not specific to digital photography: what brings about a
digital photo is the conjuncture of these tendencies that converge when the camera 'clicks'.
In particular, the idea of having subjects that are not posing. This idea is already present in
photography prior to the digital. From the 1930s, we have in fact begun to take portraits of
ordinary situations, without staging the photographic subjects. In family albums, we can notice
that the portraits, where everybody is positioned and posing for the photograph, have been
replaced with more 'natural' photos. In the field of artistic photography, we could reference
Doisneau and his 'stolen moments'. But this tendency existed even before the photo: let us
think, for example, of Caravaggio's portraits. *Saint Matthew*, for example, with his dirty foot
pointed towards the viewer, is a painting profoundly marked by this tendency.

As for the rapid acquisition of the final product – the digital photo does not need to be developed and printed –, we notice that this tendency is identifiable prior to digital photography: in
1948, for example, the Polaroid appeared. But also, before the Polaroid, one described the
skill of a painter by pointing out their speed, and their ability to capture a moment – such as
the impressionists. To these, we add a particular technique, that of the *camera obscura* which,

108 Eric Méchoulan, 'Intermédialité, ou comment penser les transmissions', *Fabula Colloques*, 2017, http://
www.fabula.org/colloques/document4278.php [accessed 29 May 2018].

if it was undeniably at the center of photographic development, also characterized painting, at lease from the Venetian Veduta. Instead of identifying digital photography as a separate medium, we could therefore concentrate on the practice of taking a photo – the very moment that the photo is taken – and analyze the ensemble of tendencies that are activated at this precise moment: this is what it means to account for *mediating conjunctures*.[109]

Servanne Monjour analyzes this complex mechanism of dialogue between techniques, discourse and practice in the context of digital photography by referring to 'remanence'.[110] The idea of remanence allows a reinterpretation of the notion of remediation as defined by Bolter and Grusin in *Understanding New Media*, all while avoiding the major pitfall of this essay, which is a certain teleological temptation that we have already referred to. The impression we have when reading the two media theoreticians (at least focusing on this landmark work) is that the remediation phenomenon is part of a globally linear process. The idea of remanence avoids this pitfall by producing unforeseeable and intertwined non-linear processes. There is therefore remanence of analogue structures in digital photography practices – such as, for example, the 'analogue' effects of digital devices –, but there is, at the same time, paradoxically, digital remanence in analogue practices. For example, the idea of the virtual image was indeed the basis of analogue photographic discourse: the negative is the virtual that is then updated through a series of different prints. And yet this virtual image seems retroactively borrowed from our ordinary digital image – notably that of the Raw image file. We are faced with an element that is established in reverse:

> Therefore, the conditions of conveying an idea participate entirely in its invention, even if it means integrating it, in reverse, into a tradition, that is, retroactively associating it with a meaning that it has perhaps not (yet) entirely had – or, in certain cases, that it did not have at all.[111]

This approach therefore allows us to avoid a teleological vision of media's progress. There is not, in other terms, a linear development of all media that would become more and more transparent, for example, as in the illustrious metaphor of the 'Black Box' that Henry Jenkins[112] denounces, as well as what we have seen in this chapter. The change of mediatic practices often occurs following an oscillation of ideas and tendencies that more or less have a strong influence on these practices according to the moment in consideration. During a certain period, for example, the notion of representation could be dominant enough to alter artistic practices. The desire of representation as an action consists in rendering present

---

109 As opposed to cultural series, mediating conjunctures insist that we account for the multiplicity of elements present at the moment of the action: there are multiple histories relating to the many different elements. One cannot include the ensemble of mediating conjunctures in a continuous series. If a link exists between cultural series and mediating conjunctures, it is that the latter can be thought of as the crossing over of different cultural series.

110 Servanne Monjour, *Mythologies postphotographiques: l'invention littéraire de l'image numérique*, Montréal: PUM, 2018 http://parcoursnumeriques-pum.ca/introduction-158.

111 Monjour, *Mythologies postphotographiques: l'invention littéraire de l'image numérique*, p. 146.

112 Henry Jenkins, *Convergence Culture: Where Old and New Media Collide*, Revised edition, New York: NYU Press, 2008, p. 13.

something that will no longer be, during a given period, one of the forces that, for example, leads to making a painting or to staging a theater play. This thirst for representation inevitably generates the development of techniques and apparatus that allow us to render present that which is not or, at least, to give us the illusion – perspective, the *camera obscura*, photography, the phonograph, etc. In their essay, *Remediation. Understanding New Media*, Bolter and Grusin in this way attempt to identify the vast historical periods wherein the principle of 'immediacy' was dominant, which we assimilate with the representational system. But they are also interested in other historical moments that they associate with hypermediacy, wherein this desire is less evident. This is the case in the 20th and 21st centuries, primarily in the visual arts field. The very concept of representation competes with other aspects of artistic experience as in, for example, life. If we think about the effects of the analogue that digital photo software offers, it is clear that the desire of representation has been replaced by other desires. Obviously, we cannot think this reduction of want of representation as progress: criticizing the idea of representation was fundamental in Plato's thinking, which did not prevent a great comeback of *mimesis* in other historical periods, as well as its decline in others. It really relates to an oscillating movement. But this movement does not, as Bolter and Grusin suggest, limit itself to two poles – namely, immediacy and hypermediacy. In moving towards representation, other tendencies arise, for example the tendency to create world hierarchies in structured taxonomies, the tendency to organize complex linear structures, the tendency to mathematize reality, or the tendency to create networks (that we observe with the Romans, builders of road networks, with the English *Railway mania* and *Canal mania*, with internet pioneers...). We could identify an ensemble of tendencies to describe particular mediating conjunctures. It is what we will do in the last chapter of this work in an attempt to justify the mediating conjunctures that crystallize today with a substantive adjective: 'the digital'.

It now remains to see how these forces and tendencies can be interpreted not only as factors that frame and shape the action, but also, and above all, as performative elements. This will be our focus in the following pages.

Performativity and Metaontology

Defining mediating conjunctures, as we have seen, can be misleading. To grasp the moving forces that characterize the moment of an action, to describe them, leads us back to giving a name to a context that eventually becomes a particular *medium*. If we are capable of identifying the spatial, symbolic or material characteristics of a given moment, we must also be capable of naming this context, of freezing it, and of making it a concept. In other words, it would be possible to say that a particular ensemble of mediating conjunctures is theater, another radio, another television, another writing on the computer... But in this way, our interest in the notion of mediating conjunctures disappears.

As we have seen, the intermedial approach tries to resolve this problem by concentrating on the dynamic that leads to the production of mediations rather than on the media itself. And yet with this approach, one still runs risk of essentializing the process. Intermediality, in fact, always runs the risk of becoming an ontology. We have shown that, in the first phase of its history, intermediality would concentrate on the idea of an in-between that in some way ends up essentializing the poles that it sought to emancipate. If intermediality is what is found between many media, from this arises an essentializing determination of these media. In the *postmediatic approach*, one puts aside the idea of medium and insists on the importance of the remediation dynamic and on what Lars Elleström calls 'media transformation'. This movement, that manifests as a result of a growing lack of differentiation between media and an erasure of the barriers that demarcate them – think about a television system and home cinema or current mobile phones and their multifunctional qualities – is precisely what highlights the centrality of mediation and medialities. It is otherwise on the basis of these medialities that the transmedial dynamic is founded: the mediatic contents are treated and distributed in accordance with specific medialities. This indicates well that, if there are no longer media, we are not in an unclear continuum where it would be impossible – even unreasonable – to identify a particular mediatic situation.[113]

If intermediality does participate in this performative shift marking the end of the 20th century and the beginning of the 21st, the idea of mediating conjunctures discussed here marks a crucial step in the development of thinking about mediation, as well as that of performativity. Intermediality is, above all, performative, regardless of the period and regardless of the system – mediatic, postmediatic – beneath which it manifests. By resorting to this idea of mediating conjunctures, we advance reflection on performativity by reversing a solidly entrenched logic. Our approach is not isolated, but part of a larger movement that we could consider 'mediationist', that varies from media theory (Galloway) to *new materialism* (De Landa, Braidotti) or to Karen Barad's *agential realism*. According to Barad, 'reality is not composed of things-in-themselves or things-behind-phenomena but "things"-in-phenomena'.[114] From this logical shift, and its considerable repercussions, emerged the major concept of intra-activity. Within the intermedial field that interests us, what we consider to be mediating conjunctures largely corresponds to the intra-active dynamic observed in agential realism. While mediatic intermediality relies on interaction, postmediatic intermediality would rely on intra-action. While the first assumes the pre-existence of predetermined independent entities to be related to one another, it is on the other hand 'through specific agential intra-actions that the boundaries and properties of the "components" of phenomena become determinate and that particular embodied concepts become meaningful'.[115]

If mediating conjunctures are always something specific, the question therefore arises: how do we consider the difference between one mediating conjuncture in relation to another without falling into a mediatic essentialism? In other words: is it possible to not abandon a discussion on the specificities of certain mediating conjunctures without adopting an essentialist approach?

113 See, for example Rosi Braidotti, *The Posthuman*, Cambridge, UK; Malden, MA, USA: Polity Press, 2013. and Manuel De Landa, *Assemblage Theory*, Speculative Realism, Edinburgh: Edinburgh University Press, 2016.

114 Karen Barad, 'Posthumanist Performativity: Toward an Understanding of How Matter Comes to Matter', *Signs: Journal of Women in Culture and Society*, 28.3 (2003), 801-31 (p. 817) https://doi.org/10.1086/345321.

115 Barad, 'Posthumanist Performativity: Toward an Understanding of How Matter Comes to Matter', p. 815.

We propose – more as a possible opening for future research than as a genuine solution – adopting an approach that could be described as 'meta-ontological'. The core principle of metaontology is that Being is always the result of the mediation process. Instead of thinking of a Being in itself which is thereafter meditated to give way to a Being for us, or a representational Being, metaontology considers mediations as original.[116] We could therefore adopt an ontological approach, which aims to grasp the specificity of mediating conjunctures, without renouncing their elusive nature. Indeed, the originality of mediations implies an original multiplicity of Beings, and therefore a multiplicity of ontologies.

Beyond falling into a unique definitional approach to ontology, which crystallizes the real into stable concepts, metaontology positions itself in ways other than that of ontology, in a way that allows us to take into account the multiplicity of ontologies. This 'meta' must not be interpreted as an overriding: metaontology takes into account the multiplicity of ontologies without presenting itself as a super-system, precisely because it recognizes the irreducibility of different ontologies. From a metaontological point of view, the real is a dynamic of entwined ontological planes. There is no essence of things, but a plurality of essences that encroach upon one another. There is no Being, but a plurality of multiple-beings. In accepting this original and irreducible multiplicity, we consider metaontology a performative discourse on the essences whose entwined dynamic constitutes as the real – or, more precisely, the reals.

According to a metaontological approach, what we say, here and now, on mediating conjunctures is one of many interactions that determine its emergence. But this emergence is always tied to a precise context that makes the essence of mediating conjunctures plural and dynamic. Metaontology is therefore a performative discourse that is here at this very instant of writing. What is produced then – this writing that is a discourse on mediating conjunctures – is metaontology. Metaontology is found in mediating conjunctures and, at the same time, it produces it. The fact that at this very moment we write, produces the very meaning of this writing. However, we are in the process of speaking of mediating conjunctures – we are in the process of trying to grasp them. But this act, as a metaontological act, is not an attempt to pin down or crystallize them: we do not define mediating conjunctures. We express them. We express them because the mediating conjunctures that we are in the process of talking about are nothing more than the ensemble that is in the process of being produced in the very instant that we are in the process of writing. These mediating conjunctures are, at the same time, the product and the condition of possibility of that which is in the process of happening while we write. Metaontology is the act of expressing, in a space detached from ontologies, that which is produced at the very moment of this discourse. At the same time, it is entirely thrown into the flow of reality and detached from the discrete time that founds the possibility of essences.

---

116 This point can be understood in line with the agential realism of Karen Barad, *Meeting the Universe Halfway: Quantum Physics and the Entanglement of Matter and Meaning*, Second Printing edition, Durham: Duke University Press Books, 2007: instead of opposing agency and structures (p. 26), Barad proposes that we consider the ontological value of these entanglements.

The metaontological approach can therefore take into account the dynamics, oscillating between performativity and identifying mediating conjunctures, and help us to character-ize mediatic practices without reducing them to fixed and essential units. This leads us to analyze the process of naturalization which presents itself as the inevitable aftermath of any performative gesture. This is covered in the following chapter.

MEDIATING CONJUNCTURES AND MEDIATION

In the previous chapter, we explained why we prefer 'mediating conjunctures' to the concept of 'medium', which still dominates media studies and communications. We propose this shift not only because of the growing difficulty of defining an individual medium but also, more generally, in response to the complexity of mediation phenomena as they manifest themselves in the present day. This is not a matter of tinkering with our conceptual apparatus to account for a more elusive reality. What is at stake is our ability to develop new tools to better grasp the scope of the mediation processes – of which we are, in one way or another, the agents – and the issues they raise.

In this chapter, we will continue our discussion of mediating conjunctures, focussing on what they produce, namely mediation *per se*. Our purpose is to show how this conceptual shift fits into a broad epistemological renewal rooted in a new way of thinking about ourselves as human beings acting in the world.

We will follow the same route as in Chapter 1, beginning with an overview of the mechanisms at work in any mediation process, as understood during the mediatic period of intermedial thinking, and then looking at the model's shortcomings and why we need to move beyond it to grasp the nature of mediation in the postmediatic age and its associated issues. We will conclude by proposing a new definition of mediation, that is, by describing what occurs when a set of mediating conjunctures is operative.

Historical Perspectives on Mediation:
Individuation, Naturalization and Sociomedialities

While resisting the essentialist temptation to posit historically specific phenomena as ahistorical constants, we cannot ignore the hold that words have on our way of being in the world. Indeed, we would say that their relative stability is, in many ways, the mainstay of community life. We communicate a fluid reality using frozen forms, a tension that parallels the paradox of flux and the photogram, the contradiction between the flow of lived reality and the limited, static representation captured by a snapshot. We include these two dimensions, the communicational and the representational, in this brief historical overview advisedly, for they are the site of the fundamental problems for the concept of mediation today.

Historically, mediation has consisted of differentiating what would otherwise be undifferentiated and indissociable from an amorphous mass or flow. Any mediation is accompanied by a process of singularization, or what we have been calling discretization. We can identify two types of discretization processes: polarization and individuation. Both are discursive in nature.

Polarization splits a segment of reality in two: it creates a polarity between two elements that are placed in opposition to form a binary structure. The segmentation of reality that it effects is clearly arbitrary and not rooted in any 'natural' necessity, for reality is continuous. Polarization is the foundation of the general laws of logic: the principles of non-contradiction and excluded middle. It cannot be day and night at once (principle of non-contradiction), for language identifies night with not-day. It's either day or night; there is no third option (principle of excluded middle).

The second form of discretization of reality is individuation. Richard Grusin, the coauthor of *Remediation: Understanding New Media*, which we have discussed[117] and to which we shall return, defines mediation as 'the process, action, or event that generates or provides the conditions for the emergence of subjects and objects, for the individuation of entities within the world'.[118]

Such entities can be constituted only if we take a section of the whole, isolate it, and consider it as a unit in and of itself. This entails drawing a precise line through the continuity of reality. Quantitative measurements are one example: a litre is a discrete part carved out of a continuous mass. The litre is 'individuated', separated from the rest. In the case of units of measurement, the demarcation is entirely arbitrary. In others, it may be based on characteristic features made salient by the process of polarization: referring to mammals focuses attention on certain characteristics of animals, to the exclusion of all others. The concept of medium arises precisely from such a process of individuation, starting from the dynamic and complex reality of practices.

By practices, we mean characteristics that are common to different actions. Agent A performs action X, agent B performs action Y. Actions X and Y are enmeshed in a series of elements: a context (C), a space (S), a time (T), a set of ideas and values (I), technological devices (D) and so forth. When one or more of these elements are the same or similar, when they converge, we can speak of a practice.

Let us say, for example, that actions X and Y share C and D. We then have a practice we may call CD. If, at 4 p.m. on April 4, Luca (L) takes a photo of Montreal from the lookout on Mount Royal (action X), and at 8 a.m. on June 20, Jean takes a photo of the Eiffel Tower from the Champ-de-Mars (action Y), there are some common features to the two actions. Similar devices were used (even if we don't know the specific cameras with which Luca and Jean took their photos); in both cases a cityscape was photographed; and so forth. We may therefore speak of a 'photographic' practice.

Clearly, practices can be organized into broader or more precise categories, depending on the elements taken into consideration. 'Vermeer paints a view of Delft in 1660' (action Z) could be said to share some elements with the two previous actions. We could then speak of a 'cityscape representation practice'. Or we could specify that Luca and Jean took their photographs with smartphones and define the practice as 'smartphone photography'.

Specifying the common thread by which media practices are defined constitutes the individuation process for that practice. Once a set of practices has been individuated, it can be

---

117 See Chapter 1.
118 Richard Grusin, 'Radical Mediation', *Critical Inquiry*, 42.1 (2015), 124-48 (pp. 137-38) https://doi. org/10.1086/682998.

assigned a name, understood and interpreted as a unit. This is the process that leads to the identification of what we call 'theater', 'radio', 'television', 'music' and 'cinema', to name just a few examples.

In this sense, the practices we include under the rubric 'cinema' are not discontinuous with the ones we call 'photography' or 'theater'. The discontinuity resides in the names we assign to them, which are produced by a process of discretization.

The arbitrary nature of these processes would be less problematic were it not for the fact that they are a way of structuring reality. As we have said, polarization and individuation are products of language. This observation takes us back to the performative dimension of language. As Austin demonstrated, language does more than convey messages; it 'does' things in the world and, in a way, creates reality. But more than this, the process of discretization also has an institutionalizing effect. Discourse not only produces the entities it names, but it endows them with an institutional existence and an ontological status. Certainly, the institutionalizing effect is not automatic: it isn't enough to name a polarity or an individuality for it to attain an institutional existence. The discourse that produces the discretization must circulate, must be shared, and must stabilize: it must always say more or less the same thing (or at least give the impression of doing so).[119]

To return to the example of units of measure, the definition of a meter is not rooted in any necessity: the same distance could be measured in inches, or leagues, or miles. But a discourse emerged, first in the scientific community, then in business practices, and eventually became the standard, in some communities, and stabilized. The institutionalization of a discretization process is bound up with specific, and yet shifting mediating conjunctures that are at once its original environment and its product. Obviously, one could propose a different unit of measure and challenge the institutionalization of the meter, but before another institutionalization process can take hold, it would have to stabilize and be shared across the community in question. Applied to media, institutionalization is the discursive process of discretization that results in the emergence of a medium.

The stabilization of discourse, which completes the institutionalization process, triggers another important phenomenon: the 'naturalization' of the discretization and its transformation into something ageless. This is accomplished when the discretization no longer needs to be justified, when it no longer appears tied to cultural variables, but instead seems to belong to the natural order of things.

---

119 Ian Hacking gave an example of the institutionalization of discourse in 2005, in his Philosophy and History of Scientific Concepts course at the Collège de France. He discussed the process by which certain medical categories (obesity, autism) become institutionalized. He argued that statistical discourse and the scientific community's definition become operational only when a group of people begin to recognize themselves in those categories. Hacking identifies at least four moments in the process: (1) classification, (2) people, (3) institutions and (4) expert knowledge and popular knowledge. His lecture is available online at https://www.college-de-france.fr/media/ian-hacking/ UPL6120975782849689510_Hacking2004_2005.pdf]{.underline}](https://www.college-de-france.fr/ media/ian-hacking/UPL6120975782849689510_Hacking2004_2005.pdf. [Retrieved on 29 May, 2018].

In the case of media, a naturalized medium becomes a fixture of its users' daily lived environment. In *When Old Technologies Were New*,[120] Carolyn Marvin provides a striking example of how a medium becomes naturalized. She describes how the telephone entered the daily lives of a steadily growing congregation of individuals (in homes, businesses and organizations of all kinds), how it changed their lives and transformed its users. She notes that it is always a small group of initiates that promotes a new technology. At first, they are the only ones to master its use and realize its utility. But unlike other types of knowledge, this 'technological literacy' does not establish a permanent elite, for it soon comes to be widely shared.[121]

At the level of the user, naturalization has the effect of blurring the distinction between the human and the non-human. Marshall McLuhan defined media as 'extensions of man',[122] but when media usage protocols are assimilated by users to the extent of becoming automatic, it is hard to say where the extension begins. Naturalization renders the mediation transparent to the user of the medium. Saying 'hello' or 'I'm hanging up', dialling a number, hearing a voice (or not) after several rings, all belong to a body of knowledge that transmutes into a set of reflexes on the part of the telephone user. These protocols might be expected to accentuate the opacity of the mediation process, but instead they create a transparent effect as a result of their assimilation by the user. The varieties of opacification discussed in Chapter 1, exemplified by the radiophonic voice that exposes the artifice of the theatrical voice, may be understood as processes of 'denaturalization' (of what had previously been naturalized).

We have been using the term 'sociomedialities' to describe these specific socialities that give rise to media practices. Clearly, the naturalization of discretization processes tends to render the sociomedial dimensions of some individual and collective behaviours commonplace or even invisible. It might be concluded that the concept is not useful, since sociomedialities inevitably collapse into socialities. But it must be borne in mind that 'new media', or new dominant trends within mediating conjunctures are always appearing and disturbing users' behaviours.

Postmediatic Perspectives

Two points that emerge from this overview of mediation are worth noting. The first is its anthropocentric nature. Everything revolves around the (human) user, of whom media are understood as an extension. The non-human, including technology, is subject to the human and serves to increase human capabilities. This anthropocentrism breeds an anthropomorphism that may well be called into question.

The second point is the pivotal position of language in this conception of mediation. Language is central to the naming of the things we singularize and to the new environment thereby

---

120 Carolyn Marvin, *When Old Technologies Were New: Thinking About Electric Communication in the Late Nineteenth Century*, Nachdr., III, 1990.

121 Marvin calls this process 'technological literacy as social currency'. Marvin, *When Old Technologies Were New: Thinking About Electric Communication in the Late Nineteenth Century*, p. 9 ff.

122 Marshall McLuhan, *Understanding Media*, New York: Signet, 1966.

created. These two tendencies, anthropocentrism and logocentrism, are at the heart of current debates in intermediality.

As we saw in Chapter 3, for mediation to exist there must be a combination of elements that we have called 'mediating conjunctures'. A notable feature of these elements, in addition to their shifting and partly random nature, is their heterogeneity, consisting, as they do, of human and non-human agents and variegated entities and phenomena. This defining characteristic of mediating conjunctures stems from their instability, their diversity and the fact that they are not exactly reproducible. To Bolter and Grusin's contention that there is no original mediation, we might add that any mediation is always new, or better still there is never a mediation that is identifiable as such, but always mediating conjunctures.

But this perpetual novelty is itself variable in scope and intensity. In a given space of time – and both time and space may be extended – the degree of novelty may be relatively weak, and mediating conjunctures may be grouped together on the basis of similarities, or what Wittgenstein called 'family resemblance'. Likenesses of this kind ground categories such as the arts, which have been distinguished from other types of mediation and also divided into subcategories, such as the temporal arts and the spatial arts, or distinguished on the basis of certain aspects of the devices they employ, such as the use of words, musical notes, drawings, etc. These entities were then named and organized into a hierarchy, which has changed over time.[123] Of course, the essentialist naming of the conjunctures not only discretizes, but also reifies them, concealing to some degree their fundamental and dynamic characteristics.

What is true of the arts applies as well to media. The term 'medium' as we refer to the it here was, in the late 19th century, both of the need to distinguish traditional representational practices – the arts – from new practices in which the non-human occupied a hitherto unknown role (thanks to the growing use of electricity and the fast-paced development of sound and image reproduction technologies). This revolution in agency, by which we mean the scope and nature of the role that various agents play in mediation processes, did not occur without friction. Resistance sprang up, nourished by explicitly anti-mechanical and technophobic discourses, often tinged with nostalgia. In the 1930s, while Henri Gouhier was seeking the 'essence' of theater (in our terms, a nexus of mediating conjunctures that is stable over time) in the live and 'real' co-presence of actor and spectator, independent of any technical mediation, Walter Benjamin was developing the concept of the aura (which can be understood as a specific quality of certain forms of mediation that weakens with the intercession of non-human agents, mechanical or technological).

Postmediatic intermediality reflects the mediation processes it describes: it not only takes note of the role of the non-human in mediation, but strives to understand how the human and non-human are intertwined, by a logic other than the subjection of one to the other. Intermedial research on contemporary theater (in which technology figures prominently in the form of reproduced images on the stage, the use of microphones, sound environments, etc.) is a good example. From the outset, the intermedial approach was distinguished by

---

123 During the Romantic period, for example, music was often considered the highest art form.

its concern with what was then called the materiality of media, and it made an important contribution to re-evaluating the role of the non-human in the development and operation of mediating conjunctures. The new value it ascribed to the non-human represented a departure from the anthropocentric conception of technology inherited from McLuhan.[124] Considerable effort was devoted to understanding the specific dynamic of the non-human and its modes of interaction with human agents. A prime example is the research on virtual reality, and on synthetic characters. At the same time, the admission of the non-human considerably expanded the already vast arena in which mediating conjunctures are produced and operate. This necessarily brings us back to the foundation of all mediation: its performative character and the nature of that performativity.

The Formless: Beyond Language

A few years before Gouhier embarked on his quest for the constants that define theater and endow it with a specific 'essence', and Benjamin began framing his intuited 'aura' in theoretical terms, the French writer Georges Bataille decried the unhappy effect of words on our perception of the world. To elude their reifying and reductive power, Bataille proposed a new kind of dictionary, one that 'would begin from the point at which it no longer rendered the meanings of words but rather their tasks'.[125]

Beyond his distrust of words, Bataille's aim was to free humans from the grip of the sign and the inseparability of signifier and signified, the cornerstone of the then-ascendant structuralist school in linguistics. Bataille refused to collapse the totality of experience into our acquaintance with the invented forms that are words and their meanings. The paradox Bataille confronted is the same as the one we have discussed in preceding chapters: the words that are supposed to describe the world mask it in the act of description. The result, complains Bataille, is a radical disjunction between the linguistic sur-reality that is supposed to represent the world and reality itself, to which language, with its limitations, cannot gain access. What interested Bataille, though, was not so much the limits of language, as reality itself, the object of his quest. He called it 'the formless', a notion he developed in the margins of and in connection with the surrealist movement. Formlessness was an attempt, then, to speak the ungraspable, and the concept opens the door to what might be called a poetics of the unexpressed.

Bataille provides no satisfactory illustration of the formless,[126] nor, of course, any clear, stable definition of the concept, which would have been a logical contradiction. But some characteristics recur in his discussions of the notion: it is heterogeneous, ungraspable and a process.

This last point is particularly interesting for the intermedial approach. It recalls Aristotle's view of matter as an undifferentiated mass that is individuated and rendered recognizable and clas-

---

124 Only insofar as McLuhan relegated the non-human to an extension of the human, as we have noted.
125 Dominic Faccini and others, 'Critical Dictionary', *October*, 60 (1992), 25-31 (p. 27) https://doi. org/10.2307/779027.
126 In general terms, he suggests the examples of the earthworm, the spider and the spittle.

sifiable by form. Bataille plainly had this in mind, as the entry for 'formless' in his *Dictionnaire critique* makes clear: 'Thus *formless* is not only an adjective with a given meaning but a term which declassifies, generally requiring that each thing take on a form'.[127] The act of declassification therefore demands both the presence and negation of form: the formless is not just the absence of form but also, at the same time and inseparably, its active suppression. From this perspective, then, the formless possesses a destructive dimension that is, paradoxically, also a gateway to something else. Bataille's dictionary would 'no longer [render] the meanings of words but rather their tasks'[128] and it was through those tasks that the reader, listener or spectator would open up to the presence of the formless and experience it. The formless contains a strong performative vein even as it calls for a redefinition of agency.

But at the heart of Bataille's liberation struggle is not so much this aspect, but rather his distrust of words. He wants, first and foremost, to wrestle himself free of the exclusive authority of the sign and the signifier/signified pairing which, in limiting and imprisoning the formless, reifies it and substitutes it for the plenitude of experience, and the company of human language, an invented form that is at once arbitrary, reductive and plural. In conceiving formlessness, Bataille not only rejects the cohesion of the sign but discredits the efficacy of all representation, viewed as autarkic, self-constructed, narcissistic artifice in the final analysis. In other words, signs (particularly but not exclusively linguistic signs) are tools for the subjugation of the non-human to the human. It is this anthropomorphic, anthropocentric dimension of form, and the self-containment it effects that Georges Didi-Huberman underscores in *La ressemblance humaine ou le gai savoir selon George Bataille*.[129]

A startlingly inventive writer though he was, Bataille was unable to describe the formless other than by what it is not. It's easy to see why: how can one say the unsayable within the bounds of language? This question marks the limits of the logical framework within which Bataille was operating, namely the logocentric outlook for which language is the privileged means of apprehending the world and all that exists that can be apprehended by language. While Bataille had the intuition that there was something beyond communication, he did not move beyond a preliminary stage of reflection, one that consisted not so much of blazing new trails towards the vast continent of the formless, as of dismantling the routes that do not lead there. His enterprise heralds the great debates that would attend the rise and fall of semiotics – and of linguistics in general – but its lasting importance resides mainly in this fundamental and fraught question, precisely because it is counterintuitive: what is the undifferentiated? To ask this question also involves asking how the undifferentiated is to be mediated, regardless of whether it was always undifferentiated or became so when detached from human language.

These questions raised by Bataille nearly a century ago are at the center of today's intermedial debates and they are key to uncovering the workings of the mediating conjunctures at the heart of our discussion here. The concepts of 'radical mediation' and 'excommunication' that have been advanced recently, and the issues surrounding the naturalization of mediation pro-

---

127 Faccini and others, 'Critical Dictionary', p. 27.
128 Faccini and others, 'Critical Dictionary', p. 27.
129 Georges Didi-Huberman, *La ressemblance informe: ou le gai savoir visuel selon Georges Bataille*, 2019.

cesses and their artefacts that we discuss here, attest to the currency of Bataille's questions today and to the counterintuitive character of postmediatic intermediality.

Radical Mediation: Updating the Remediation Model

In Chapter 1, we looked at some limitations of the remediation model and showed how other, non-transformational dynamics could connect mediating practices such as hypermediality and transmediality. Independently pursuing his investigation of mediation processes that had begun with Jay D. Bolter, Richard Grusin has returned to the remediation model in order to clarify some of its basic features. Notwithstanding the modesty of this claim, the changes he has proposed to the model are substantial and contribute to a renewed conception of mediation, which Bataille would have undoubtedly welcomed. Grusin's reconsideration of the key remedial concepts of transparency, immediacy, opacity and hypermediacy produced the essay *Premediation: Affect and Mediality After 9/11* in 2010 and then the substantive article 'Radical Mediation' in 2015.

During the 16 years between the publication of *Remediation: Understanding New Media* and 'Radical Mediation', Grusin disentangled the concepts of transparency and immediacy that had been bound together in the category 'transparent immediacy'. This is a turning point, for with this move Grusin frees mediation from the representational yoke to which he and Bolter had originally tied it to. This liberated mediation is what he calls 'radical mediation'. And Grusin goes further still. He argues that immediacy is limited neither to the products of mediation, nor to its erasure – *i.e.*, its transparency – and, what's more, that it is mediation itself that is immediate: 'Mediation is not opposed to immediacy but rather is itself immediate'.[130]

Grusin returns – or rather relegates – the two-pronged concept of transparent immediacy to an instance of 'media correlationism', which might be regarded as an updated version of the historical notion of a copy's fidelity to the original. Likening transparent immediacy to a theory that originates in the speculative realist school is a momentous shift that marks a radical turn in Grusin's thinking. More notably still, Grusin expands the definition of mediation from an in-between to a general process of differentiation of all things in the world, human and non-human.

How does this major amendment to the remediation model by one of its authors help advance thinking about mediating conjunctures? In two ways. First, it tacks towards performativity. Grusin's 'immediate mediation' now includes what might be called 'intransitive mediation' – that is, mediation that essentially refers to nothing beyond what occurs at the moment of mediation: no model is transmitted, no pre-existing before or elsewhere is summoned in the here and now of the mediation.

Secondly, and paradoxically, this development leads to an impasse that, while not invalidating the remediation model, limits its scope and, most importantly, forces us to rethink mediation. In solving one problem, the model's confinement within the realms of representational logic,

---

130 Grusin, 'Radical Mediation', p. 129.

Grusin creates another, one that is highly instructive. He says: 'Where remediation focused largely on the visual aspects of mediation, radical mediation would take into account the entire human sensorium'.[131] This passage and others from the landmark article suggest that, like Bolter, Grusin has trouble conceiving visual mediation outside of representational logic, which is why he entirely abandons the concept of opacity – it is never mentioned in the article – and also why he concludes that immediacy makes the concept of transparency inoperative and useless. However, stating that mediation is immediate is insufficient for defining it. By excluding the interplay between transparency and opacity from the scope of his inquiry, Grusin forgoes an essential analytic tool.

Let us compare a tightrope walker and a magic show. In both cases, unless the performance contains an overlaid narrative that refers to something beyond itself, the mediation is intransitive: the circus performer walks the tightrope, the magician performs his trick. It all happens in the here and now of the show. But there is a basic difference between the two: we assume there is no trick to the high-wire act, and yet we constantly seek one in the magic show, as we know it exists. What, then, do we see in these performances? In the first, we see everything; in the second, everything except the thing we're looking for, even though it's there. The magic show is a paradoxical show of not showing: we see that we don't see. So this is not opacity as it is understood in the classic representational paradigm, an opacity that obscures, masks or, in Zola's terms, 'distorts' that which ought to be represented transparently. In this case, nothing is being re-presented: there is only live presence. This show of not showing, this 'aesthetics of the impossible' as Darwin Ortiz calls it,[132] rests on partial, willful masking of basic elements of the mediation. Even as it reveals itself, the specific mediation of the magic show eludes the spectator.

Continuing the comparison, we can say that, in the high-wire act, the mediation is intransitive, immediate and transparent, since we see everything there is to see. Or at least that is how we experience it. In the magic show, the mediation is more subtle and complex: we can see that we don't see, which is why we cannot speak of opacity in this case, for in a magic show as in any show, we cannot see nothing. If there is neither opacity nor transparency, as we are not within a representational logic, what are we dealing with? We are dealing with an opacity that is not opaque and which, paradoxically, lets us see that we don't see. The process that unfolds in the magic show might be called 'inopaque'. Inopacity is, in a way, the operative concept in the aesthetics of the impossible which Ortiz refers to.

The concept of radical mediation is of considerable heuristic value. Situating immediacy within mediation itself, as Grusin suggests, rather than in the product of the mediation, is a major historical departure. If mediation consists in endowing objects in the world with an anthropomorphic aspect by re-presenting them, radical mediation establishes a different, more direct or immediate relationship with the world, which could be called formless since it does not rest on form.

---

131 Grusin, 'Radical Mediation', p. 132.
132 Darwin Ortiz, *The Annotated Erdnase*, Pasadena, Calif: Magical Publications, 1991.

It is unfortunate, however, that Grusin has exchanged one indissoluble pair (transparent immediacy) for another (the immediacy of mediation), thereby excluding the possibility of other combinations. While it was necessary to situate mediation within a wider logic than that of representation, the persistence of the representational mode of mediation in all its forms, particularly in the arts and performance, is something that we must accommodate for.

Mediation as Excommunication

Alexander R. Galloway was one of the fiercest critics of Bolter and Grusin's remediation model and a prolific proponent of abandoning the concept of medium in favor of mediation. Today, the majority of intermedialists subscribe to his formula: 'not media, but mediation'. In 2014, he published *Excommunication: Three Inquiries in Media and Mediation*, co-authored with two other New York media theoreticians, Eugene Thacker and McKenzie Wark. The book opens with an apparently innocuous question that directly echoes Bataille's reflections: 'Does everything that exists, exist to be presented and represented, to be mediated and remediated, to be communicated and translated?'[133] A series of examples from various spheres leads them quickly and predictably to answer in the negative, which raises a second question: how, then, do we know that something exists? And a third: how can something still be transmitted; in other words, how is it that something is mediated?

Rehashing their argument in detail would be irrelevant here, but it unquestionably represents another significant advance in intermedial thinking about mediation and effects a major paradigm shift: while Grusin has released mediation from the representational cage to which he and Bolter had confined it, he still keeps it penned within the perimeter of communication – a larger enclosure, to be sure, but an enclosure nonetheless. Galloway and his co-authors go further by proposing to remove mediation from the communicative framework. Just as Grusin showed that the representational schema is only one remediation model among many, Galloway et al. suggest that communication is only one mode of mediation; it is neither the principle nor the matrix of mediation: 'We aim, therefore, to craft not so much a theory of mediation in terms of communication, for which there already exist a number of exemplars, but a theory of mediation as excommunication'.[134]

What, then, is this 'mediation as excommunication' that operates outside communication, *i.e.*, outside strictly human communication (as it has been conceived until now), and that encompasses the human and non-human without distinction? It is 'a double movement in which the communicational imperative is expressed, and expressed as the impossibility of communication'.[135] What they are referring to is not the absence of a message or an inter-

133 *Excommunication: Three Inquiries in Media and Mediation*, ed. by Alexander R. Galloway, Eugene
     Thacker, and McKenzie Wark, Trios, Chicago; London: The University of Chicago Press, 2014, p. 10.
134 Galloway, Thacker, and Wark, *Excommunication: Three Inquiries in Media and Mediation*, p. 11.
135 Eugene Thacker, 'Dark Media', in *Excommunication: Three Inquiries in Media and Mediation*, ed. by
     Alexander R. Galloway, McKenzie Wark, and Eugene Thacker, Trios, Chicago; London: The University of
     Chicago Press, 2014, p. 80.

rupted communication; neither is it a performative failure of mediation, since something, they insist, is still being mediated. We don't know what, but we sense it, just as, in a magic show, we see that we don't see, without (theoretically) knowing what it is we're not seeing.

What the three authors of *Excommunication* are trying to articulate corresponds, in many respects, to Bataille's project. As we know, media theory originally revolved around what was thought to be a universal communicative process, the best-known model being Shannon and Weaver's classic scheme.[136] Their 1948 model is based on a one-way, linear, structural conception of communication. It breaks down (we would say 'discretizes') processes into sequences, including encoding and decoding, and the agents, human and non-human, who perform them. The model follows the 'message' through successive stages, from the 'information source' and the 'sender' to the 'receiver'.

The proposal Galloway and his colleagues make does not overcome the difficulty Bataille encountered in defining the formless and modeling its dynamics. This does not, however, vitiate its value, for it in fact unveils a new stage in debates about intermediality and mediation. Just as we can regard mediation as excommunication – that is, understand it outside the communicative framework – we can speak of excommunicating mediating conjunctures – *i.e.*, instances of mediation that are located outside the classic model and hence cannot be explained by it. Thacker calls these types of mediation 'Dark Media'. It might be suggested that the magic show, with its 'inopacity', is a salient example. The function of what might be called 'dark mediating conjunctures', which rub shoulders with more traditional mediating conjunctures, no longer aims to make the inaccessible accessible or to connect what is separate: 'Instead, media reveal inaccessibility in and of itself – they make accessible the inaccessible – in its inaccessibility'.[137]

Conclusion

Georges Bataille forcefully defended the idea of a pervasive formlessness from which humans have cut themselves off by the very means they use to mediate it. They seek other paths, but find none to this formlessness untouched by any discretization, for they remain imprisoned in a logocentric logic. What they glimpse remains within the realm of communication and can be expressed and transmitted only by their means.

The slow process of maturation that led Richard Grusin to a 'radical' form of mediation is the most important change in the theory of remediation to date. Breaking with the deep-rooted and widespread notion that the function of media, like that of the arts, is to 'represent', an idea that undergirds the remediation model, Grusin abandons the representational paradigm and casts mediation as experience in the first instance. In this case, what counts is the immediate, the mediation, not its product.

---

136 John R. Pierce, *An Introduction to Information Theory: Symbols, Signals & Noise*, 2nd, rev. ed, New York: Dover Publications, 1980.
137 Thacker, 'Dark Media', p. 96.

Excommunication is the most recent development in thinking about mediation: communication is one form of mediation but not the only one, nor the most effective. Grusin has emancipated the varieties of mediation from the confines of representation; Galloway, Thacker and Wark have situated them beyond the limits of communication, in the region where Georges Bataille sought the formless.

Conceiving of mediating conjunctures as other than representational practices that translate the world into human language, accepting the inaccessible in its inaccessibility and experiencing it as such, constitutes a decisive double break in the development of what we call postmediatic intermediality, and describes a new way of being in the world and living together in and with the world.

THE DIGITAL AND THEATER AS MEDIATING CONJUNCTURES

In this final chapter, we would like to show how intermedial thought, as we conceive of it, is able to renew our approach to certain questions which are, or have been, at the heart of important debates in various research fields. We limit this study to two examples which seem to us particularly significant for our study. Firstly, we will focus on the digital in order to show how intermediality can help us to better understand this reality, its nature and its conjunctures. We will then reorient the discussion toward theater, a site of revaluation and major reforms throughout the Long 20th century (from 1880 to today). We will therefore focus on two concepts that have occupied, and continue to occupy, a considerable place in theatrical reflection: presence and theatricality.

The Digital as Mediating Conjunctures

The Digital does not Exist

The word 'digital' has invaded the public sphere for about twenty years. We use it in everyday language, we talk about it a great deal in the media; for institutions – governments, universities, businesses – it is as a key word. The word had first been used as an adjective: 'digital document', 'digital communities', 'digital identities', 'digital economy', 'digital environment' [...] up until 'digital culture'. More recently, it has become a substantive: it is referred to as 'the digital' and even an 'object of analysis'.[138] Enthusiasm for the digital has otherwise contributed to the renewal of Media Theory: the emergence of what could be called the 'digital turn' has shaped, in all disciplines, a new and general interest in theoretical questions that have, until now, seemed confined to the fields of Information Sciences or Communication Sciences. The emergence of discourse relating to the digital has been accompanied by the reappearance of a rhetoric on revolution, a rhetoric that is by no means unprecedented nor exclusively tied to the digital turn, as we will see.

In the following pages, we would like to analyze the digital with the theoretical tools offered in this book. This exercise will allow us to simultaneously test the operational value of the various concepts that we have developed, understand the context that they have emerged from and provide an interpretation of a discourse and a series of practices that characterize our time. All of us are, indeed, buried in 'digital culture',[139] and the changes we reference with this expression are among the reasons that justify the theoretical need for this book.

From the emergence of the term, which was related to technology and its explosive use in the 1990s, we have envisaged the digital in many ways, but each kind of discourse on this subject can be considered an attempt at a discretization that aims to institutionalize not yet

---

138 Patrik Svensson, 'Envisioning the Digital Humanities', 6.1 (2012) http://www.digitalhumanities.org/dhq/vol/6/1/000112/000112.html [accessed 5 December 2014].
139 For the definition of this expression, see Milad Doueihi, *Digital Cultures*, Cambridge, Mass.: Harvard University Press, 2011.

adopted practices. In other terms, the digital is nothing other than the name by which we try to naturalize practices whose characteristics are not sufficiently clear to avoid entering into other adopted institutional categories. We can assert that there is not, even today, a discourse on the digital that is stable enough to consider it as a point of reference. Many, sometimes contradictory, parallel debates circulate in public space, none of which seem to generally converge in order to overtake others and thereby emerge as institutional discourse. It would otherwise be impossible – precisely because of the instability and the non-institutionalized nature of these debates – to analyze them, or even further, to simply establish an exhaustive list.

We will here take into account three ways of describing the digital that seem to us the most present and therefore the most pertinent for analytical purposes: the digital as a parallel space – the debate surrounding virtual reality –, as a medium or an ensemble of media – the debate on the age of communication – and lastly as an ensemble of revolutionary devices – the debate on 'New Technologies'. These three approaches do not really correspond to modes of critical or scholarly analysis, but they represent rather the collective imaginary surrounding the digital, the way of understanding this phenomenon which is particularly visible in mediatic and commercial discourse and which has a considerable impact on the general perception of technological changes.

The Digital as a Parallel Space

One of the debates on the digital turn includes that which tends to identify the emergence of a space parallel to our primary space, a space determined by different rules and that, often, competes with this primary space.

Offering a historical record of this debate is a particularly delicate task because, evidently, it does not constitute an unprecedented *topos*: it had been present on numerous occasions throughout the history of thought, and not only in the last century. Indeed, we can recover this debate on a parallel and unreal space in the case of television – as with Baudrillard[140], for example, or with Popper[141] – but also that of the knighthood novels of the 17th century – that we see for example in *Don Quixote*. But we could also go much further in retracing the history of this *topos* from the myth of Plato's Cave, a myth that presents exactly the same logical structure: many parallel spaces whose degrees of reality are different, yet difficult to identify.

We know, therefore, that his debate does not apply purely to the digital, as it constitutes rather a reactualized *topos* enabling us to understand particular practices. More precisely, we can say that this debate asserts itself in relation to the digital in the 1980s, with the appearance of the expression 'virtual reality'. Many texts have already traced this history,[142] which we will very briefly summarize here. Virtual reality is a modelled, immersive and interactive space

---

140 Jean Baudrillard, *Le crime parfait*, Paris: Editions Galilée, 1995.
141 Karl R. Popper and Giancarlo Bosetti, *The Lesson of This Century: With Two Talks on Freedom and the Democratic State*, London; New York: Routledge, 2000.
142 We deem it relevant in this instance to cite Marcello Vitali Rosati, *S'orienter dans le virtuel*, Cultures numériques, ISSN 2118-1926, Paris, France: Hermann, 2012.

produced with the help of a machine. The idea is that, thanks to computers, we can produce a parallel reality that is completely artificial and completely 'realistic'. This idea is more
an imaginary space rather than a technical reality: indeed, this concept has considerably
developed in science fiction tales, in literature – Gibson's *Neuromancer*, for example – and
even more so in cinema – from *The Matrix* to *Existenz*. Even if the technical possibilities for
materializing this fantasy continue to develop and if the notion of virtual reality takes on a
fundamental importance in industry discourse, notably since Oculus Rift's commercialization
in 2016, this concept does not allow us to sufficiently describe the digital turn. We could say
that if, in the 1980s with the invention of the concept of virtual reality by Jaron Lanier, this
model seems to need to be the dominant model in computer science development, today it is
clear that virtual reality concerns a very limited number of experiences among the multitude
of information technology uses.

The parallel world *topos* is promising and interesting when we analyze its recreational or
fictional use. It was, in fact, what Janet Murray observed even in the 1990s with her book
Hamlet on the Holodeck.[143] But this notion becomes insufficient and problematic in understanding a series of practices that are in no way fictional, nor recreational, and that take on
a large importance in our daily use of digital tools. In other terms: in the 1980s and 1990s,
one would have thought that recreational use would constitute a privileged use of emerging
digital tools. As practices have developed, we have realized that gaming is only one of many
possibilities relating to these tools, and probably not the most interesting.

Taking the place of the debate on parallel, virtual, fictive, fluid and navigable space – a series
of terms tied to parallel space –, arises a debate on concrete, hybrid and inhabitable space.
The digital is always considered a space, but a space that integrates with non-digital space by
blurring the difference between physical and non-physical which characterize the first debate,
as well as the one relating to the fictive and non-fictive. We can here notice an initial phase
of naturalization: the notion of discontinuity, which would characterize virtual reality rhetoric,
is replaced by that of continuity. The digital turn appears to be not so much a revolution, but
rather something natural that has always been a part of social practices.

The Digital as Medium

Another way of envisioning the digital turn has been to think about all tools – or even groups
of tools – as a medium or as an ensemble of media. This debate is very complex and differentiated, as well as difficult to convincingly summarize, but we can try to understand it starting
with its most extreme formulation, which consists in speaking of a digital medium as though
an ensemble of technologies were united to give way to a very powerful medium that would
become the only device of any mediation form. This is the idea behind Jenkins' *Black Box*.[144]

---

143 Janet Horowitz Murray, *Hamlet on the Holodeck: The Future of Narrative in Cyberspace*, Cambridge,
      Mass: MIT Press, 1998.
144 Henry Jenkins, *Convergence Culture: Where Old and New Media Collide*, Revised edition, New York:
      NYU Press, 2008, p. 13.

This approach is tied to a series of debates that have developed, since the 1960s, in the field of Media Studies and in the field of Information and Communication Sciences. From this point of view, one tends to think of media as devices enabling communication between two poles – whether these poles be either living beings, machines or even conceptual structures. One therein develops models to define the communication process – as an exchange of information between two poles – and one tries to understand media as an in-between that, at the same time, renders possible and structures the communication. This debate is often accompanied by the principle of transparency and immediacy – and remains closely tied to a representational regime. One presupposes that there would be an 'ideal' communication wherein there would be no 'information loss' and that therefore there would be a perfect medium which is completely 'transparent'.

We have seen the different problems that arise with this approach and the different ways of perfecting this debate to avoid such downfalls. It is the history of the intermedial mediatic period, with its essentializing biases, that are increasingly overrun by theories that attempt to free the representational paradigm's exclusiveness and the polarities that it presupposes.

To think the digital turn in terms of media poses a large number of problems and it is probably these problems that have contributed to the changes relating to theories on media. This is what we have seen with the different intermedial postmediatic approaches and in particular with the critique of the communication paradigm developed in *Excommunication.*[145] Indeed, the very notion of communication is tested in the digital context because, unless we accept an extremely wide definition of communication, it becomes impossible to think in terms of communication about phenomena, such as the algorithmic gathering and analysis of private information for the purposes of profiling, algorithmic organization of content as in the case of a search engine or the use of cartographic tools to find a restaurant.

'New' Technologies

Rhetoric relating to novelty does not, in reality, constitute an isolated and independent debate. It is more a general way of envisaging the digital, a way that also affects the two other debates mentioned here and that, in addition, is used to refer to many non-standard practices well beyond the digital turn. It is a debate that tends to identify elements that break with a hypothetical 'tradition'. Applied to the digital, this debate is based on the key notion of 'new technologies', whose paradox is easy to guess. Novelty can only be an ephemeral quality and can therefore not be acknowledged on a definitional basis. The adjective 'new' is therefore unrelated to the thing that it would like to characterize. Every technology has been 'new' during a certain historical moment within a certain community – as provocatively stated by the title of Carolyn Marvin's book *When Old Technologies Were New.*[146] Naming a technology as 'new', on the one hand, emphasizes a presupposed break between this technology and

---

145 *Excommunication: Three Inquiries in Media and Mediation*, ed. by Alexander R. Galloway, Eugene Thacker, and McKenzie Wark, Trios, Chicago ; London: The University of Chicago Press, 2014.

146 Carolyn Marvin, *When Old Technologies Were New: Thinking About Electric Communication in the Late Nineteenth Century*, Nachdr., III, 1990.

the preceding one and, on the other hand, labels all that came before this novelty as obsolete, participating in a somewhat teleological discourse (which is a very good – and very easy – sale argument for businesses).

Regarding the notion of a presupposed break: the idea of novelty refers, from a theoretical point of view, to the possibility of clearly and cleanly separating one chronological moment from another. This exercise is obviously always characterized by an arbitrary aspect, as the caesura between two chronological moments – as far as it can be justified – is never without ambiguity and is, to an even lesser extent, based on consensus. The difficulty for historians to delineate historical periods serves as proof of this. In addition, the argument of novelty becomes weak when we consider the fact that it has been widely used for more than a century: in the field of media for the telephone, the radio,[147] etc., and in art, for example the avant-garde.

The notion of novelty has become an omnipresent *topos* in commercial discourse that serves to establish the rapid obsolescence of products, pushing users to buy the 'new' version of the same object as soon as it is available. If we rely in part on analyses of this discourse, novelty is very often identified by two elements indicating 'progress': speed and simplicity. Novelty is, in other words, always understood in commercial discourse as an advancing of a teleological process that moves toward dreamlike perfection consisting in a technological device's complete transparency. It is transparency that we criticize in this book. The products are labelled with higher numbers indicating better versions (iPhone 5, 6, 7, for example) and each version 'exceeds' the preceding version often by improving the speed and offering increased simplicity.

This ideology is clearly justified by the necessity to sell 'new' products: it is, however, equally necessary to distance ourselves from it if we hope to gain a theoretical understanding of the digital turn.

Modern, Contemporary... Digital

The debate on novelty allows us to better identify the origin of the need to make such common use of the word 'digital'. We maintain the theory that this word does not derive from a real revolutionary change that presents itself as a veritable chronological break – that it be produced with the arrival of the web in the 1990s or prior. No break disrupts temporal or historical linearity. Our history is continuous, it is its institutions that discretely alter it. In other terms, our practices change in time, and we modify our customs, our technologies, our way of relating to the world. And this change is continuous. It must also not be considered an evolution, as it does not move toward 'progress'; it is not teleological. Quite simply, our practices change. The institution standardizes practices because they are based on a state of things at a given moment. From the moment the institution crystallizes its standards, practices continue to change, until the moment that the institutions can no longer standardize them, as these practices have become too different. It is at this moment that the institution must

---

147 Micheline Cambron and Marilou St-Pierre, 'Presse et ondes radiophoniques', *Sens Public*, 2016 http://www.sens-public.org/article1199.html [accessed 29 May 2018].

change. And this change is done immediately – with a change of standards that can only be discrete. It is at this moment that we invent new words to describe the change. And it is at this moment that we begin to refer to revolution. There is therefore revolution – or at least a discrete and important change –, but of the institution and not of the practices.

Let us consider some examples. The institutional devices that standardize editorial practices were put in place during the 18th century.[148] It is interesting to note that more than two centuries separate Gutenberg's invention of the creation of institutions. It is in the 18th century that copyright was put in place, that practices were standardized, that editing became a framed practice.[149] Publishing companies became institutions. The states created laws to regulate these practices. The institutions enabled standardization and gave meaning to these practices – but also gave them a name. And yet, since the 18th century, these practices have progressively changed. The meaning given to words has changed, the importance attributed to the author has changed – for example in the 1960s[150] – techniques have changed, as have readership, the number of literate readers, economic and social conditions, etc. These changes create unease from within the institution, as these norms no longer correspond to the realities of these practices: practices distance themselves more and more from the way the institution represents and regulates them. It is at this moment that crisis unfolds and the need for institutional change emerges. The symptom of this crisis and of this need for change is the word digital, with which we name the disparity between real practices and those practices that are standardized by the institution. In this sense, the expression 'digital editing' names all the continuous changes that separate 18th century editing from today's editorial practices. With the expression 'digital editing', institutions try to take further stock of the state of practices in order to be capable of once again standardizing them so as to continue to manage them. We are living a moment of institutionalization.

This same debate applies to a large number of institutions: if we think of personal identity management, of teaching, of research, of art, of communication. In the case, for example, of creating university positions in 'digital literature', we witness a change in practices that require the creation of this label. We speak of 'literature' to refer to a certain number of practices that have crystallized over years and centuries. These practices evolve up until a time where the difference between what we mean by literature and what real practices are becomes too important to not pose a problem. The university does not know what to do with practices that do not enter into the institutional framework: they therefore begin to use the word 'digital'. In this sense, the digital has nothing – or very little – to do with computers. It may be revealed that information technology has played a role in the change of our practices, but this is not the primary role, and above all not the only one. This is why, in the end, digital literature does not refer to literature made by computers, but rather literature during the historical moment of computers. The expression 'digital literature' could be replaced by 'today's literature'. These

---

148 See Elizabeth L Eisentein, *The Printing Revolution in Early Modern Europe*, Cambridge: Cambridge University Press, 1983.
149 See, for example Mark Rose, *Authors and Owners: The Invention of Copyright*, Cambridge, Mass.: Harvard University Press, 1993.
150 Roland Barthes, *Le bruissement de la langue*, Seuil, 1993, Michel Foucault, 'Qu'est-ce qu'un auteur ?', in *Dits et écrits*, Paris: Gallimard, 1994.

are, in other words, practices that probably relate to literature but that diverge too much from institutionalized practices for us to simply consider them 'contemporary literature'.

And yet this brings up a series of questions relating to the standardization process of practices, on the role of institutions, on their place in society. Institutions are, by nature, always behind the times. Their standardization is based on something that is always, by definition, outdated. Practices are always beyond foundational standards, they are always laden with an institutionalizing character that is creative and normative. Each time someone creates literature, they do something that is not really what we mean by literature but that redefines the meaning of 'literature'. And this creativity undermines institutional standards. Each time, the institution tries to catch up by changing. This change is what we can call 'institutionalization' or 'standardization' our even 'naturalization', as we have seen in the previous chapter. In the case of digital literature, for example, institutionalization consists in creating prizes for 'digital literature', classes, associations, editors that make 'digital literature' and so forth (we could also refer to the work done for a number of years by the *Electronic Literature Organization*). This institutionalization implies that new practices no longer seem foreign and deviant. They become normal. This also implies that norms are appearing. We begin to know what 'digital literature' is and therefore what it 'should' be. Prior to being institutionalized, alternative practices were by no means standard: they were simply practices that did not conform to institutional definitions. But after institutionalization, they begin to be standardized, regulated: we know what we must do if we want to reclaim the sector. This implies that creative and innovative practices will end up by moving away from a given definition and will again find themselves marginal to the institution.

The standardization phase is, of course, necessary for being able to organize our society. And yet, there are two ways of standardizing practices: one is diachronic and the other synchronous. This means that we can try to standardize a very limited number of practices over a very long period (diachronic) or a very high number of practices over a very short period (synchronous). Let us think about the definition of practices that we assemble under the name of theater. We observe that human beings in many societies and for many centuries do something particular and very specific: they display their actions. We notice that very often this display of actions is done within a pre-defined setting, we notice the social role of this display of actions and the ensemble of (technical, cultural, anthropological) devices that structure these practices. All of this, we call 'theater'. With this word, we also try to define what the theater is; we try to structure it. We try to understand it and to standardize it so that we may manage its role in society. It is based on this institutionalization that we can create theaters, systems to subsidize theater productions, theater teaching programs, and so forth. The same reasoning applies to literature, cinema, television, radio. These are diachronic definitions of very precise practices. Obviously, institutionalization provides a snapshot: it disregards the differences between a heterogenous ensemble of practices in order to place them in the same group and name them. One could provide a degree of tolerance regarding the more or less important differences in order to assemble a more or less elevated number of practices (for example, by saying 'cultural activities', we group 'literary', 'theatrical', 'cinematographic' practices, etc.).

Synchronous institutionalization consists in taking a very heightened number of practices relating to a very short time frame. In this way, we very often refer to what is known as 'culture', for example the culture of a period. Humanism, as a culture that characterizes the 15th and 16th centuries, assembles a series of very heterogenous practices that are nonetheless characterized by a common denominator, a kind of 'spirit of times'. We could say that religious cultures also enter into this type of institutionalization even if, in the case of religion, we group together a synchronous vision and a diachronic vision. But from a certain viewpoint, we could assert that it is important is to first standardize a very heightened number of practices in the present moment. What counts is that all of today's Christians behave in the same way in a determined situation. The fact that a hundred years ago things were different does not count for much. Of course, the daily actions of a Christian from a hundred years ago were very different from the actions of a Christian today, simply because the changing environment, social concerns, etc., were not the same. But the fact of being Christian today alters the ensemble of possible practices (one is Christian when one eats, when one makes art, when one works, etc.).

The digital tends to function in the same way – it is Milad Doueihi's central theory in his book *Digital Cultures*,[151] wherein the author rightly makes comparisons between the digital and religion. With the word 'digital', we define a very heterogenous ensemble of practices that, as its common denominator, takes place in our present historical moment. In using this word, we tend to converge all these practices, even if they are very different: posting a photo on Facebook or writing a scholarly article, purchasing an airplane ticket to go on vacation or live-tweeting during a rock concert... This convergence can be seen as a characteristic of contemporary culture. This is true and false at the same time. It is true from an institutional point of view: discourse on the digital tends to understand a very heterogeneous ensemble of practices as a unit, and it thereby converges them. In institutional discourse, a 'digital' expert will be consulted in relation to purchasing strategies for Google servers, just as he would be for psychological problems arising among adolescents using Facebook, or for the latest Houellebecq novel in which the author plagiarizes Wikipedia. What is derived from many different disciplines today converges into one realm: the digital.

In reality, the practices themselves converge today no more than they would have converged one hundred or two million years ago. It is just our way of naming and understanding them from an institutional point of view that has changed.

This specification can be understood in an essentialist way: there exists a certain number of practices whose essence is to belong to the same group. For example, 'digital' practices or 'theatrical' practices. But this specification could otherwise be understood in a dynamic way: it is precisely what we propose with the notion 'mediating conjunctures'.

The Digital as Mediating Conjunctures

If the word digital allows us to *synchronously* unite very heterogenous practices, and if this group indeed serves to highlight the divergence of current practices in relation to the ensemble of institutionalized practices, it should be possible to identify – even in a non-exhaustive

---

151   Doueihi, *Digital Cultures.*

way – the reoccurring features of these actions in order to extricate the 'mediating conjunctures' that are the foundation of this digital turn. In other terms, it should be possible to identify
a series of forces, tendencies, values, ideas, agents that dynamically interlace to give way to
current conjunctures that we adopt into common usage with the label 'digital'.

It is necessary to again return to the arbitrary character of this operation of identification. As
we have said, mediating conjunctures are always plural and multiple, and every act aiming
to identify them involves an arbitrary aspect. Nevertheless, we can better grasp this common
discourse by specifying the forces and the tendencies that seem to be present in the current
context to give way to this idea of digital culture that Milad Doueihi refers to.

We propose to focus our attention on four aspects of the digital which seem to us particularly
pertinent: the notion of a network, the notion of performativity, the notion of non-linearity and
lastly the idea of the mathematization of the world.

Network Culture

The word 'digital' is strongly associated with the development first of the internet and then
of the web. Digital culture is undoubtedly a network culture, as we can see by the large
number of scholarly works dedicated to the subject.[152] Ideas such as contribution, sharing,
connectivity, collaboration, etc., which are so present in today's culture, bring into question
many institutional gains – tied to the idea of originality, for example, or of the individual, or
again of the author – that are founded on a network structure which seems to us increasingly
present. Clearly, this large presence of networks and the impact that these structures have
on our societies are not the only features of the digital world. In other words, the digital is
without a doubt a network culture, but it is not the only one, nor the first. The strength and
omnipresence of the network are some of the digital turn's particularities, the tendencies
in play in the mediating conjunctures that we call 'digital', but they are not a revolutionary
element nor in themselves unprecedented. Retracing the history of network culture would be
impossible, as it has crossed our societies' entire history as we remember, for example, the
road networks put in place by the Romans – a centralized network, of course, but one that
aimed to facilitate the possibility of decentralization, a possibility on which the empire was
thereafter founded. We could cite other situations where the idea of a network has played a
fundamental role: the constitution of the posting network from the Middle Ages, *Canal mania*
and then *Railway mania* in 18th and 19th century England... Network culture has therefore
developed for millennia and their technical, economic and cultural capabilities, as well as
their interpretative and symbolic capabilities relating to the network's consequences, which
acquire over time, are present during the development of informational networks, the Internet
being the latest example. In other terms, the notion of the network has always existed, even
if it were able to have a differing influence over the centuries. Digital mediating conjunctures
renew this notion, but they are far from inventing it *ex nihilo*. In addition, understanding this

152 See, for example Geert Lovink, *Dark Fiber: Tracking Critical Internet Culture*, Electronic Culture – History,
     Theory, Practice, Cambridge, Mass: MIT Press, 2002, Doueihi, *Digital Cultures*, Tiziana Terranova,
     Network Culture: Politics for the Information Age, London; Ann Arbor, MI: Pluto Press, 2004.

long history of network culture is necessary for correctly analyzing the debates surrounding current technologies: informational networks have benefited from the infrastructures that have previously been created for other networks: notably the telegraph and railways. Optical fibre, for example, crossed the United States following the railway, as railway societies are the only ones to have been given permission to bury a cable across the entire width of a country in avoiding telecommunication societies by needing to ask permission from millions of different land owners. Therefore, the geopolitical issues tied to the construction of east-west railways in the 19th century have an impact today on Internet geopolitics.[153]

The Idea of Performativity

Another characteristic of the digital, or even, another force in play in the establishment of the mediating conjunctures that we call 'digital', leads us to the timeless opposition between two paradigms relating to our interpretation of reality: a representational paradigm and a performative – or if preferred, operational – paradigm. To put it simply, we can think about writing practices just as we think of forms of representations of the real, that is, a way to form a relationship between a signifier – writing – and a referent – the real, to which writing refers to. In this sense, we can think about writing as having a mimetic function and insert it into the series of representational forms, such as have been practiced and analyzed since Antiquity – from Plato to today. Within this context, the question to be asked would relate to the way linguistic acts correspond to their referents or, otherwise stated, the relationship between truth and fiction: is writing a faithful representation of the real or is it a fiction, such as the novel?

And yet digital culture does not seem to privilege this paradigm and instead suggests another: the performative paradigm. We have studied the notion of performativity, and the effervescence around this term shows that this paradigm acquires an importance more and more central to the way we think the world.[154]

To briefly explain, we could say that according to the performative paradigm, writing must be considered as an act producing the real. In other words, it is not a case of representing a reality, but of constructing it. Digital culture privileges this approach. When we write something on the web – and here we imply, by writing, all forms that produce a trace, and therefore also the fact of publishing a photo – we do so with an operational aim. We do not represent something, but rather we do something – on the web I buy plane tickets, I organize my bank account, I work, I buy a book, etc. Even when I write a review for a restaurant on Tripadvisor,

---

153 See, for example Gill Plimmer and Daniel Thomas, 'Network Rail's Fibre Optic Network Attracts Telecoms Interest', 2016 https://www.ft.com/content/173b6c06-f1da-11e5-aff5-19b4e253664a [accessed 10 April 2018]. or Jane Tanner, 'New Life for Old Railroads; What Better Place to Lay Miles of Fiber Optic Cable', *The New York Times*, 2000 https://www.nytimes.com/2000/05/06/business/new-life-for-old-railroads-what-better-place-to-lay-miles-of-fiber-optic-cable.html [accessed 10 April 2018].
154 On the topicality of the performative paradigm in relation to the representational paradigm, see also Karen Barad, *Meeting the Universe Halfway: Quantum Physics and the Entanglement of Matter and Meaning*, Second Printing edition, Durham: Duke University Press Books, 2007 who affirms that this change also interests the physical, which passes from the representational paradigm of Newton's physics to the performative paradigm of quantum mechanics.

I do this so I can do something: for example, to change the ranking of the restaurant in question or its visibility. In fact, I am in the process of producing the restaurant, of contributing to what this restaurant is – for example, a good or a bad restaurant, a fish or a meat restaurant.

And yet, once again, it is easy to demonstrate that this tendency to privilege the performative paradigm, as a characteristic of our digital culture, was not recently invented. The two paradigms have always existed and the importance of one or the other, as well as their prevalence among common receptiveness, have not ceased to vary over the course of history, as we see in the differing conceptions of mimesis between Aristotle and Plato. For the former, mimesis has an exclusively representative function. For the latter, it functions as a structuring of the real. The mediating conjunctures that we crystallize using the word digital are, in fact, characterized by a pre-eminence of the performative paradigm, but this pre-eminence is only one of the forces in play in current mediating conjunctures.

Non-Linearity

Non-linearity is among the characteristic tendencies of the digital, as it, too, was obviously not born alongside the development of information technologies. The example of the Alexandrine manuscripts is particularly illuminating in this regard. It is enough to cast a quick look at a manuscript such as the *Palatine Anthology* to realize this. This is a 10th century codex containing a collection of Greek epigrams. The manuscript is particularly meaningful because, thanks to it, we have access to a large part of the Greek poetry that we know today: the manuscript is therefore of fundamental cultural importance.

Many characteristics of the manuscript seem interesting given our current discussion: the fact that it is an anthology, the fact that it is the result of the sedimentation of many textual layers – it contains poems written between the 7th century BC to the 10th century AD, the fact that it contains fragments and that it proposes different non-linear classification systems. All these characteristics are specific to digital writing. The idea of the anthology, which developed during the Alexandrine age in the 3rd century BC, consists in composing collections that are at the same time exhaustive – they give an overall idea of a literary form or of particular themes – and partial – as it is always the result of a selection. The *Palatine Anthology* is the recompilation of ancient Alexandrine anthologies – in particular that of Meleager. Obviously, the anthology implies a fragmentary form, and the fragmentary form invites a non-linear structure. The question that herein arises is as follows: how to produce a non-linear form in using a paper support that is suited to linear writing? Compilers and copyists have found many solutions to this challenge: one of them is the use of scholia, which are comments in the margins that often create relationships between the different fragments. The fragments are organized in thematic order (the love epigrams, the death epigrams, the religious epigrams, etc.) and the comments are placed in the margins to identify the epigrams of the same author or those that speak of the same person or the same city. In this way, the scholia offer a non-linear organizational form. The *Palatine Anthology* seems to reveal the great 'desire' of the web to also characterize past centuries. Technological developments are probably in part determined by this desire to orient and structure them. We can try to retrace this penchant for non-linearity by citing, for example, the ideas of Vannevar Bush, and then of Ted Nelson, to make a link between a principle already present in the 3rd century BC and the structuring of the web as conceived of in the 1990s. The hyperlink, the layering of different versions, systematic rewriting – plagiarism, pastiche, parody – have always existed and the digital has only absorbed these culturally ancient tendencies.

Mathematization

Another tendency of 'digital' mediating conjunctures can be seen in the more or less explicit dream that has haunted our minds for many millennia. One sees it here and there in Egyptian lists, in Aristotelian catalogues, in the mnemonic rules of Florentine Neoplatonists of the Renaissance, in Leibniz's mathematical models, in the affirmations by the big names of the web: the world is made up of an enormous mass of information, the knowledge and cultivation of which could enable an almost complete mastery. It would therefore be possible to know everything, to foresee everything, to do everything. And there are two purely human limitations that inhibit the retention and cultivation of this entirety of information: accessibility and calculability.

Not all information is entirely accessible to humans, even if we consider that information exists, that it lies somewhere – contained in a book, retained by a group of specialists or simply observable in nature – or that it is disseminated, hidden or even beyond understanding, and this is because of an inability wherein we find ourselves decoding and standardizing languages, coding, formats. The structural problem of the library derives from this observation. The library is a device that attempts to knock down one of two limits of the dream of complete knowledge: accessibility. In a library, all the information – or at least a lot of information – is accessible. But, even if complete accessibility were achieved, one other limit would remain: the concatenation of data, calculability. In other terms, once a person has access to a large mass of information, how can they cultivate it? Having access to millions of books, I can only make use of the information that I am capable of reading and retaining. In order to cultivate this information in its totality, the information would need to make up an object calculable by a huge machine that would automatically process them. The mathematization of sciences is an attempt to resolve this second problem.

With the enthusiasm stirred by the development of the web, such a dream was born and promised both complete accessibility as well as calculability. The machine became a Leibnizian god, capable of knowing all and calculating all and promising complete accessibility and complete calculability which means complete knowledge. Tim Berners Lee can be considered a public figure who promotes this dream of complete knowledge. In a 2009 TED Talk,[155] he presents this idea in an enthused and enthusiastic manner. 'A few years ago, I asked you to put your documents online and you did so. It's great! Today I ask you to put your data on the web'. Accessibility and calculability. The web renders an enormous mass of documents accessible. A question herein arises relating to the cultivation of this data. A human cannot use this data because of its size. But if this information becomes pure data, machines can therefore understand and calculate it: complete knowledge is therefore possible. 'Raw Data

---

155 Available here: https://www.ted.com/talks/tim_berners_lee_on_the_next_web

Now!' cries Tim Berners Lee. Raw data is a formalization of the information that renders knowledge usable by a machine. The analysis of data can take place thanks to the strength of the computer's calculations. The data is pure, open to all possible associations of meaning and a machine can tie it to an ensemble of other data. The world can be known in a complete, absolute and objective way.

It is therefore evident that the idea of the mathematization of the world, which is sometimes presented as a novelty of the digital age, has traversed the history of thought.

Conclusion

The ensemble of these ideas and tendencies combines today by giving way to conjunctures whose singular and novel aspect is not found in each of these ideas considered individually, but in their concentration in shared space. It is the combination of these ideas – as well as a series of others that may please us to identify and specify – within a particular technological and economic context that gives way to the digital turn. And yet, as we have highlighted, mediating conjunctures are not stable and cannot be 'identified'. The gesture of identification that we here propose is therefore necessarily arbitrary, as it consists in isolating certain characteristics in relation to others, and to purposefully 'isolate' our observational range. As in the case of a school of fish, we can limit our gaze to what occurs within the geometric shape of the school, or we can also see how it interacts with one or many external elements. And again, we can choose a more or less extended period of observation – synchronous or diachronic. In the same way, the digital turn is a way of demarcating a field of observation within our current society by prioritizing certain aspects. In reality, behind this crystallization we find a moving ensemble of forces and tendencies for which we can always retrace a continuous line of development.

Theater as Mediating Conjunctures

The theater offers an observational field so fertile for intermedial studies that it is surprising that a delay of nearly twenty years separates the emergence of the first intermedial concepts (in the middle of the 1980s) and their application to dramatic and dramaturgic practices (in the middle of the 2000s).[156] theater, which has witnessed remarkable mutations thanks to electricity, and which experiences an important renewal attributable to the breaking down of traditional disciplinary barriers and to the invasion of digital technologies, can be considered, as Chiel Kattenbelt suggests, as the ideal intermedial practice.[157] It would therefore prove logical that theater studies integrate central intermedial concepts into their heuristic arsenal. As a dynamic and fertile practice where, since its beginnings, arts and technologies cross and combine, and where the concept of mediation is crucial, the theater finds itself in fact

---

156 Freda Chapple and others, *Intermediality in Theatre and Performance*, Amsterdam; New York: Rodopi, 2006.

157 Chiel Kattenbelt, 'Theatre as the Art of the Performer and the Stage of Intermediality', in *Intermediality in Theatre and Performance*, ed. by Freda Chapple and Chiel Kattenbelt, Amsterdam; New York: Rodopi, 2006, p. 37.

tied, in one way or another, to all the great mediatic and technological upheavals that have shaped the world of communications and entertainment for over a century and that are at the heart of intermedial theory.

This surprisingly belated penetration is not the outcome of chance or of random heuristic negligence, it is largely explained by what we have called a 'mediatic resistance' occurring within the world of theater. The intermedialists affirm the primacy of relationships over the objects caught within the dynamic and the genesis of media practices and, above all, they place at the center of their preoccupations the question of materiality and the role of technology – the non-human. In doing so, they have done little to draw sympathy from a community who, for more than a century, have forged their identity based on the real-life actor (their voice, their body) and their 'direct' relationship, reputed as pure, true and natural, with another real-life person, the spectator. By rejecting essence (including that of theater) and binary oppositions – for example, the live versus the mediated – onto which dominant theatrical discourse is constructed, intermediality calls into question certain foundations of theater studies and invites us to redefine them. But this meeting also concerns circumstantial factors. Intermedial thinking was deployed at a moment when theater studies had hardly begun to emerge as a separate (university) discipline. It was therefore vital, in order for theater studies to assert its legitimacy, to distance itself from literary studies, a discipline from which they sought to free themselves.

We now approach two subjects, largely discussed within theater studies, that illustrate the difficulty of understanding a stage-based and dramaturgic practice in full bloom. The intermedial perspective that we adopt allows us not only to measure the distance separating the stage's reality from theoretical discourse, to grasp its inadequacy, but also to identify some of its underlying causes. The notion of presence (and liveness) and that of theatricality thereby reveal a profound crisis which has not yet entirely subsided.

Presence and its Founding Trio: Liveness, Ephemeral, Nowness

Jonathan Sterne,[158] a figurehead of American Sound Studies, recalls the slogan-concepts 'fidelity' and 'high-fidelity' that have been hammered into the sound reproduction industry[159] from the end of the 19th century up until the end of the 20th century and that, relying on mimetic logic, have no other objective than to frustrate the 'traitor' accusations directed towards its 'devices', which disrupt the link between the body of the artist and the ear of the listener. These devices break the tie that its defenders claim as 'direct', 'pure', 'natural', 'unique' alongside, obviously, any other evocation of nostalgia and anticipation of the future. Sound reproduction technologies supposedly damage the *aura*[160] of the artist and the stage that had been, up until then, the only place of the consecration of 'beautiful voices'. In addition to questions of reproduction and authenticity that it gives rise to, the intrusion of

---

158 Jonathan Sterne, *The Audible Past: Cultural Origins of Sound Reproduction*, Durham [N.C.: Duke university press, 2003, p. 215 and ff.

159 Intended as either long-distance transmission or recording, they were electric or acoustic.

160 The term, popularized by Walter Benjamin's momentous essay, was obviously not used at the time, but similar concerns were nonetheless raised.

'mediated' sound – that is to say, transmitted by microphones, loud speakers and phonographs (records) – onto the theater stage was therefore perceived and presented as a double assault, by detractors who were also the defenders of the so-called 'secular' stage: the assault of the 'new decadent culture' of mechanic entertainment against the 'other' culture which is hundreds of years old; we refer to technology's assault against the artist and against art, often reduced to that of the false against the true. Not only did this reproduced sound establish an initial historical break between the voice and the body,[161] moving away from what nature had connected, but it launched the 'unnatural' practice of acousmatic listening.[162]

This debate, which has left many still very tangible traces, in addition to what we will cover here, aimed to preserve a 'live' relationship that would be glorified for its essential, and transcendent value relating to the concept of liveness,[163] a value which, from the outset must, in the spirit of its promoters, be situated at the hierarchical pinnacle of representational practices 'seen in action' (as in cinema, records, radio or television). Nobody today contests the fact that the theater is effectively a practice above all branded by living presence – the co-presence of actor and spectator – but the hierarchical conception of practices that were attached to this argument rest on the existence of stability and of the media frontiers that intermediality reduces to discursive constructions. However, from an intermedial point of view, the most important problem lies elsewhere: it derives from the oppositions and amalgams which form the foundation of thought relating to the concept of liveness. In defining this school of thought in opposition to the mediated, one ends up considering the live and the mediated as irreconcilable, destined to mutually exclude one another. As for the amalgams, the effects were equally pernicious. Here we will retain only that which has allowed a widening of the concept of liveness to that of 'nowness', a process that irreducibly associates presence with the ephemeral and confining it as such.

This discussion purports only to highlight and to valorize that which, for more than two millennia, created essence and belonged to its episteme. But these arguments fit, in a much more prosaic way, within market-based principles. We therefore notice that, in Arthur Pougin's monumental 1885 dictionary on theater and its associated arts, which is among the first

---

161   Mladen Dolar, *A Voice and Nothing More*, Cambridge, Mass: The MIT Press, 2006, pp. 9-12.
162   Referring to Pythagoras's famous lessons given to his disciples, wherein he separated himself from them by a curtain so that they heard him without seeing him, Pierre Schaeffer developed the concept of acousmatic sound by demonstrating the role played by sight in relation to hearing. (see Pierre Schaeffer, *Traité Des Objets Musicaux: Essai Interdisciplines*, Paris: Seuil, 2002, p. 91 and ff).
163   This concept is difficult to pin down, firstly because it is complex, but also because it has not ceased to evolve. For the purposes of this study, we will limit ourselves to the definition offered by Philip Auslander: 'In theatrical parlance, *presence* usually refers either to the relationship between actor and audience – the actor as manifestation before an audience – or, more specifically, to the actor's psychophysical attractiveness to the audience, a concept related to that of *charisma*. Concepts of presence are grounded in notions of actorly representation; the actor's presence is often thought to derive from her embodiment of, or even possession by, the character defined in a play text, from the (re)presentation of self through the mediation of character, or, in the Artaudian/Grotowskian/Beckian line of thought, from the archetypal psychic impulses accessible through the actor's physicality, Philip Auslander, *Presence and Resistance: Postmodernism and Cultural Politics in Contemporary American Performance*, Annotated edition edition, Ann Arbor: University of Michigan Press, 1994, p. 37.

encyclopedic works on theater, presence is in no way referenced. Pougin does, however, methodically examine the semantic field of the word 'theater', from which he draws more than half a dozen meanings. None of them deal with presence, be it in relation to the actor or to the spectator. Pougin also dedicates a long and particularly insightful paragraph to 'theater art', but again, without alluding to the notion of presence.

> Here, a word whose significance is great, a word that represents and implies an en-semble of very diverse qualities, whose happy reunion only can enable the attainment of the entirely relative perfection allotted to human nature. Theater art is a particularly complex art, very varied in its expression as in its means, speaking at once to the spirit, the imagination, the ears and the eyes, and thereby producing impressions of a rare strength and a surprising intensity.[164]

These 'very diverse qualities' and this 'particularly complex art, very varied in its expression' grasps, without naming it, a theatrical concept that we will later discuss. If, in this definition of theater art, we see that it resonates with Richard Wagner's[165] Total Artwork project, the symbol-ists' *synesthesia dream*, or even Chiel Kattenbelt's hypermedium model, we find not a minute trace of this notion of presence so frequently evoked following the appearance of cinema!

The intermedial approach has demonstrated that there are ontological issues tied to every technological innovation – or every intermedial transfer (between an ensemble of mediating conjunctures and another) – and that these concerns become major and therefore terribly worrying when innovation has the ability to overturn usage protocols and established values. The more anxiety thrives, the stronger the media resistance that nourishes it becomes. This was the case during the electrical era, and it remains perhaps even more so in our current historical moment. Technologies founded on the digital are supple and have a transforma-tional potential clearly superior to that of electric technologies. It is therefore unsurprising that the anxieties induced by electric media are not only transferred, but amplified by essentializing discourse.

Presence and Reproduction

Pougin published his dictionary even while sound reproduction technologies (telephone, microphone, phonograph) were still in their infancy, when cinema was no more than a fan-tasy. But this is not the case for Walter Benjamin's *The Work of Art in the Age of Mechanical Reproduction*, published during the aforementioned period. He dedicates a long passage to the theater, largely repeating the criticisms and fears expressed by the playwright and director Luigi Pirandello regarding cinema in his novel *Si gira*.[166] Choosing a camera operator as his protagonist, Pirandello demonstrates, in this work the little consideration, if not contempt, that

---

164 Arthur Pougin, *Dictionnaire Historique Et Pittoresque Du Théâtre Et Des Arts Qui s'y Rattachent: Poétique, Musique, Danse, Pantomime, Décor, Costume, Machinerie [...] Fètes Publiques, Réjouissa*, Forgotten Books, 2018, p. 63, our translation.

165 The influence of the concept of *Gesamtkunstwerk* is defended by Wagner and adopted by Pougin's contemporary symbolists.

166 For which the English title is *Shoot!* (translated in 1927 by C. K. Scott Moncrieff).

he has for this 'new entertainment'. He therefore sees cinema as neither more nor less than the end of the actor's art. The cinema as anti-theater!

> The film actor, wrote Pirandello, feels as if in exile – exiled not only from the stage but also from himself. With a vague sense of discomfort he feels inexplicable emptiness: his body loses its corporeality, it evaporates, it is deprived of reality, life, voice, and the noises caused by his moving about [...].[167]

Or again,

> The artistic performance of a stage actor is definitely presented to the public by the actor in person; that of the screen actor, however, is presented by a camera [...].[168]

The 'immediate' presence of the actor in relation to the spectator, and their interaction, to which Pougin made no allusion forty years earlier, therein emerges as THE distinctive quality of theater performance and seems to guarantee its ontological superiority in regard to 'reproduced' performances. The depreciative effect extends not only to electric media, but to all forms of mediations realized with the help of technological (or non-human) devices – be they acoustic or electric.

We could undoubtedly reproach Benjamin for having granted an overly large space – an entire space, in fact – to Pirandello in the 1910s, the years of silent film, and of preoccupying himself with the writer during this period of his reflection on the *aura*, which concerns theater. We know that Benjamin defended the idea of wider sharing and better accessibility of artistic productions, if only at the cost of losing their auratic quality. If Pirandello's opinion is particularly interesting to us, it is because it indeed reflects a largely widespread feeling in the theatrical milieu of the period. But this was neither a unanimous point of view nor a progressive one. In the post-war days, Gabriele D'Annunzio, Pirandello's great rival, as well as the futurists, did not cease, on the contrary, to boast the virtue and potential of cinema. In addition, Pirandello himself did not take too long to rally to their side, performing a remarkable 'conversion',[169] to borrow Fabio Andreazza's ironic term.

This concept of theater's presence must in fact be understood as an application of the *hic et nunc* principle of the work of art, developed by Benjamin, to theatrical reality,[170] a *hic et nunc* that creates its own uniqueness: 'The presence of the original is the prerequisite to the concept of authenticity'.[171] In an era where competition between other media drives theater

---

167 Walter Benjamin, *Illuminations: Essays and Reflections*, ed. by Hannah Arendt, trans. by Harry Zohn, New York: Schocken Books, 1969, p. 229.

168 Benjamin, *Illuminations: Essays and Reflections*, p. 228.

169 In particular, when he sees an opportunity to adapt his play *Six Characters in Search of an Author* to the screen. On this topic, see Fabio Andreazza, 'La conversion de Pirandello au cinéma', *Actes de la recherche en sciences sociales*, 161-162.1 (2006), 32 (pp. 32-41) https://doi.org/10.3917/arss.161.0032.

170 And also covered in his essays *Petite histoire de la photographie* and *Sur quelques motifs baudelairiens*.

171 Benjamin, *Illuminations: Essays and Reflections*, p. 220.

to justify its very existence via a discourse on identity[172] that relies on its 'difference', the concern is simple: every disparity in relation to this 'authenticity' must be perceived as an impoverishment. In fact, the 'situations into which the product of mechanical reproduction can be brought may not touch the actual work of art, yet the quality of its presence is always depreciated'.[173] It is obviously unlikely that the defenders of theater's 'essence' had read Benjamin and were inspired by the passages that he dedicated to Pirandello. We should instead notice that these preoccupations were well and truly circulating in this troubled time and that, for the defenders of presence, art is human, and therefore all non-human intervention – technology, as it happens – impoverishes it or, worse, betrays it. But contrary to Benjamin, the defenders of theater's presence cling to a dualist model, forgetting the advantages of reproduction that Benjamin nonetheless highlights. They cling to this simplistic logic where the 'reproduced' – that which corresponds to the mediated – suffers an unavoidable loss in relation to the original. They present theater performance as one of the ultimate spaces of resistance to the domination of mechanical reproduction. This theory sometimes seems supported by Benjamin himself.

Any thorough study proves that there is indeed no greater contrast than that of the stage play to a work of art that is completely subject to or, like the film, founded in, mechanical reproduction.[174]

Benjamin not only recognized theater's particular status at the hour of its reproduction by technology, he highlighted the essential function played by the actor. It is to the actor that theater owes its originality, even its superiority. Quoting perceptual psychologist and film theorist Rudolph Arnheim, Benjamin recalls that 'the latest trend [in film consists in] treating the actor as a stage prop chosen for its characteristics and [...] inserted at the proper place'.[175]

Pirandello's complaint of the film cameras 'denaturing' interference is the same that motivated accusations of sound reproduction technologies' betrayal, as Jonathan Sterne mentions.

One could be tempted to believe that three quarters of a century later, the reality of the practice – which was not undermined by the invasion of reproduction technologies and intermedial transfers – would have relegated this essentialist discourse, along with the notion of presence, which it is based on, to the ranks of a historically categorized argument. But it is far from that. In an important work published in 1993, Peggy Phelan, a noteworthy figure in Performance Studies, persistently recalls this feature that she presents as a primordial ontological fact for theater.

---

172 The period of large definitive businesses and this identity quest is not specific to theater, as argued in a number of consequential essays by Jean-Paul Sartre (*Qu'est-ce que la littérature?*) and André Bazin (*Qu'est-ce que le cinéma?*).

173 Benjamin, *Illuminations*, p. 221.

174 Benjamin, *Illuminations*, pp. 229-30.

175 Benjamin, *Illuminations*, p. 230. (quotation from Rudolf Arnheim, *Film Als Kunst.*, Frankfurt am Main: Suhrkamp, 2002, p. 176).

Only life is in the present. Performance cannot be saved, recorded, documented, or otherwise participate in the circulation of representations of representations: once it does so, it becomes something other than performance.

One cannot counter this statement: a recorded theater representation is not the original representation. One is, however, less partial to the ideology that underpins it and that Phelan renders more explicit when she couples her declaration with a warning against all media contamination of the stage, which is, once again, not unrelated to the accusations of 'betrayal' evoked by Sterne, or the 'collapse' that Pirandello denounces.

To the degree that performance attempts to enter the economy of reproduction, it betrays and lessens the promise of its own ontology.[176]

Therefore, associating with 'the economy of reproduction' in the middle of the 1990s while the digital wave unfolded, remained a betrayal of theater's 'ontological promise!'.

Phelan's declaration would result in an aggressive reaction from Philip Auslander[177] and trigger a long and passionate debate within Performance and Theater Studies. Auslander's foundational argument was that, not only is presence not the opposite of the meditated, but it is its product, a product which is, for that matter, not exclusive to theater. Auslander's arguments were irrefutable and should have blown away all resistance, but the conflict did not limit itself to rational logic. Rather, it focused on a fundamental and hugely complex dimension: identity.

In his essay on memory, history, and forgetting,[178] published at nearly the same time as Bolter and Grusin's work on remediation and as Auslander's work on liveness, Paul Ricœur furthers his reflection on the construction of identity and on related concerns initiated about ten years prior.[179] By returning to concepts such as 'sameness' and 'ipseity', the latter corresponding more to the defenders' arguments – including Phelan – for maintaining the original purity of the practice, Ricœur underlines the flaws relating to the process of affirming one's identity and the downward spiral that this can cause. Identity is not entirely certain! 'What constitutes the fragility of identity? It is identity's purely presumptive, alleged, reputed character'.[180] It is in order to legitimize this presumption-allegation-assertion that the defenders of identity – namely, of ipseity – recall with such insistence a past that is quite obviously a reinterpreted one.

The heart of the problem is the mobilization of memory in the service of the quest, the appeal, the demand for identity.

---

176 Peggy Phelan, *Unmarked: The Politics of Performance*, 1 édition, London; New York: Routledge, 1993, p. 146.
177 Philip Auslander, *Liveness: Performance in a Mediatized Culture*, London; New York: Routledge, 2011.
178 Paul Ricoeur, *Memory, History, Forgetting*, trans. by Kathleen Blamey and David Pellauer, New edition édition, Chicago, Ill.: University of Chicago Press, 2006.
179 Paul Ricoeur, *Soi-même comme un autre*, POINTS, 2015.
180 Ricoeur, *Memory, History, Forgetting*, p. 81.

[...]

As the primary cause of the fragility of identity we must cite its difficult relation to time; this is a primary difficulty that, precisely justifies the recourse to memory as the temporal component of identity, in conjunction with the evaluation of the present and the projection of the future.[181]

As a secondary cause of this fragility, Ricœur points to alterity, 'the confrontation with others, felt to be a threat. It is a fact that the other, because other, comes to be perceived as a danger for one's own identity'.[182] The other is, namely, as we have seen, the cinema and other representational practices, but also that which, from these practices, infiltrates the theater and breaks, in the eyes of some, its 'ontological promise'.

But this secondary reason manifests itself above all in the present, whereas the former has an effect over time, both past and future: past because we shape a historical narrative made to measure for this identity, in erasing tendencies or major historical facts that alter it and magnify more secondary ones that justify it; future because identity is also an affirmation of the self, projected into the future to ensure that the projection's effectiveness depends upon the archives that one generation bequeaths to the one following. The archival gesture is not, however, mechanical nor insignificant. It allows one document, rather than another, to transcend time. Let us take two examples, one from the past, one from the present.

The 20th century historical narrative has rendered theater the 'residue of pure and authentic culture in a world of mass-media and television daftness', as Peter Boenisch rightly claims.[183] From Jacques Copeau's 'bare stage' and Peter Brook's 'empty space' to Jerzy Grotowski's 'poor theater', the narrative insists heavily on minimalist practices. These practices are often over-documented and over-commented in relation to many other experiences which are, however, not lacking in interest nor resonance in stage practice more generally. We highlight the case of the remarkable collection for the *Association de la Régie Théâtrale*[184] uncovered by Françoise Pélisson-Karro.[185] This fund includes a very rich collection of manuscript notes detailing the direction of nearly two thousand theater productions created in Paris between 1850 and 1950. Interest in these documents is indisputable, as they demonstrate well how the creators and producers of the period, in their large majority, integrated technology, including sound and image reproduction technology, into their show, without appearing distressed by ontological concerns. And yet these precious archives failed to be lost, despite being rejected by large institutions, including the National Library of France (BnF). They wandered for years, only surviving thanks to the dedication of some, before ending up at

181 Ricoeur, *Memory, History, Forgetting*, p. 81.
182 Ricoeur, *Memory, History, Forgetting*, p. 81.
183 Peter Boenish, 'Mediation Unfinished: Choreographing Intermediality in Contemporary Dance Performance', in *Intermediality in Theatre and Performance*, ed. by Freda Chapple and Chiel Kattenbelt, Amsterdam; New York: Rodopi, 2006, pp. 151-66 (p. 103).
184 Association of theater directors, stage managers and stage directors (ART) in France.
185 Pelisson, *Regie Theatrale et Mise En Scene*, Villeneuve d'Ascq, France: Presses Universitaires du Septentrion, 2014.

the *Bibliothèque Historique de la Ville de Paris* where they were finally appreciated for their cultural wealth.

We could believe that the opposition between presence and the mediated, as well as its influence on the archival gesture, would progressively ease at a time when the use of individual microphones (and the murmuring voices that it enables) would nearly go unnoticed, where the recourse to soundtracks and to videographic projections is clearly naturalized and where certain noteworthy shows, nonetheless perceived as theater shows, include no living actor on stage.[186] While this huge entry into the dreaded 'economy of representations of representations' does not seem a problem for audiences, nor for practitioners, the opposition between presence and the mediated persists. We truly perceive this within theater schools who continue to focus their education on the direct physical relationship between the stage and the room (rare are those who ensure the mastery of a role with a mic, or before a camera) but, even further, in the treatment of theater archives resulting in the direct recording of the representation.

The Archival Status of Sound and Audiovisual Recording

Certain archives seem 'natural' and possess a distinct value in virtue of an authenticity regime which, we notice, also rests primarily on the cult of presence. Theater texts, manuscript notes, photographs, programs, posters, sometimes even elements of set design or costumes, technical and administrative documents, correspondence between creative participants, critical testimonies, all of these at first seem archivable, and researchers generally have access to these documents without too much difficulty. This is not the case for the direct sound recordings of representations (by cassette or reel-to-reel recorder) that began to be used in the 1950s, nor for direct video recordings that became widespread in the 2000s (thanks to the progress and accessibility of digital technology). Audio recordings are straight out neglected, the majority of them poorly stored and currently inaudible. Very often, they are simply discarded. More recent video recordings are difficult to access. The reticence of theater companies and creative teams to make them available is well known, especially when these recordings are not the object of any post-production editing or filtering and are made with the help of a single mounted camera.

For an intermedialist, the reticence to retain these archives, which nevertheless brings us even closer to representation, is only explained because these archives, as opposed to others, are the result of a remediation of representation in its duration. They therefore weaken a foundation of the concept of theater presence, and of its ontology. In the wake of Peggy Phelan's striking essay, Gay McAuley thereby reminds us that:

> Theatre, by its nature, is an art of the present moment, and the theatre artists focus on the present of the lived experience. Performance is unrepeatable and is fascinating to performers and audiences precisely because it is unique and ephemeral.[187]

---

186 As in Maurice Maeterlinck's *The Blind* by Denis Marleau (2002).
187 Gay McAuley, 'The Video Documentation of Theatrical Performance', *New Theatre Quarterly*, 10.38 (1994), 183-94 (p. 184) https://doi.org/10.1017/S0266464X00000348.

McAuley's declaration reminds us of another made by Malcolm Cowley in 1962, at the time of the first non-radiophonic and non-phonographic audio recordings (namely magnetic recordings): 'The supremacy of the theater derives from the fact that it is always now on the stage'.[188] This fusion of the ephemeral, uniqueness and non-repetitiveness associated with the flagship concept of presence, not only goes against the human 'desire to retain',[189] but also defies all archival logic. In this simplistic reasoning where presence is opposed to the mediated, it is hard to conceive of the ephemeral other than in its immediacy, therefore in its continual disappearance, further highlighting, once again, the disconnect between discourse on theater and reality which must, on the contrary, take into account the current moment as well as the future. From there, the fall in favor of long-standing archives, from there also the response of archivists clearly summarized by Matthew Reason: 'if we do not document performance it disappears; we document performance to stop it disappearing'.[190] But no-one, aside from the defenders of essentialist discourse, claims that the end of this disappearance constitutes the persistence of performance.

We can assert, without huge risk of error, that propagators of thought on presence have turned away, for ideological purposes, a characteristic that theater shares with other practices. Instead of recognizing that presence, even in theater's most reduced form, is always the product of mediating conjunctures, they have raised this presence to the status of an absolute, insisting on its rarity and its immediacy. It is difficult to separate, in their long and complex action, that which comes from a *technophobic* sentiment, religious convictions – and the belief in natural (divine) superiority over the artificial (human) – from commercialism, nostalgia, fear, blindness or simple naivety. But let us dream for a moment: what remarkable and precious information could provide us with a video recording of the creation of *Hamlet* in around 1600 or the *Imaginary Invalid* of 1673? What an experience of a lifetime, what happiness! To what or to whom would this 'reproduction economy' harm? Surely not the history of theater? Who would be betrayed? No-one. And what would be betrayed? Nothing.

The steps taken these last years in different countries which attempt to preserve these recordings and to render them recognized learning materials shows that mentalities change and that the narrative of auratic and unique presence is losing influence. From an intermedial point of view, we can only rejoice.

Crisis Symptoms: from Theatricality to Performativity to Theatrically

The chaotic emergence and evolution of the concept of theatricality give us another occasion to test the intermedial approach in theater. In the pages that follow, we do not offer a continuous history of this major concept in theater studies, but we will concentrate on its transitional moments that reveal intermedial dimensions. This perspective is therefore slightly different

---

188 *Writers at Work: The Paris Review Interviews: First Series*, ed. by Malcolm Cowley, New York: Penguin Books, 1977, p. 100.

189 M Reason, *Documentation, Disappearance and the Representation of Live Performance.*, Place of publication not identified: Palgrave Macmillan, 2016, p. 21.

190 Reason, *Documentation, Disappearance and the Representation of Live Performance*, p. 21.

from that which we have adopted in the case of presence. Up until now, we have analyzed facts and a discourse in relation to the intermedial grid. Here, we sketch an intermedial history of theatricality.

As a composite practice in which human agents from different domains – authors, scenographers, musicians, actors, sound creators, lighting creators, performers, technicians, etc. – collaborate alongside non-human agents of an equally diverse nature, the theater is historically founded on the principle of mimesis – as are other arts, such as in Western literature – that is to say on the principle of the representation of a certain reality that is not exclusively limited to the visible universe and to everything that inhabits or has inhabited it, but to that which could have been, or should be. From Aristotle's *Poetics* to Paul Ricœur's *Time and Narrative* (1983), or to the cardinal rule of classical similitude, from the concern for the romantic 'true detail' to realist requirements and Stanislavki's 'reliving of an experience' of the 19th and 20th centuries, we have not ceased to redefine, to adapt, to draw out the concept of mimesis, all without bringing into question the existence of the triad that founds it or the invariable chronology of the mimetic action that requires, in this order (1) a source – 'the original', (2) the action to imitate and (3) the 'copy' which is its outcome.

Ricœur, in centring his reflection on mimesis in historical as well as fictional narrative, has in fact renewed and confirmed this more than two-thousand-year-old conception: 'We therefore follow the destiny of a prefigured time and a refigured time by the mediation of a configured time'. The model that Ricœur develops does, however, merit underlining the centrality of the action that he rightly describes as 'mediation'. It is the constitutive triad of mimesis and the logical which binds these elements that have progressively led to the concept of theatricality. But associated notions of action or of mediation remind us that mimesis is a dynamic process and that within it there is a central dimension which is by nature performative.

We have insisted on the fact that, all while rejecting essentialism and the idea of the existence of related irreducible invariables, we recognize that mediating conjunctures are differentiated at once from their nature and their effects. It is in this sense that we understand theatricality: it is the mediality of theater, that is to say that quality emerges from singular conjunctures that have the effect of producing a theater show or its components in a given spatio-temporal and cultural context. The processes, elements, conditions and agents required for the theater show to take place vary from one period to another, from one region to another, from one institutional framework to another, from one culture to another. Theatricality continuously evolves. And this relative instability gives rise to questions of an ontological nature that challenge institutional discourse. In fact, this instability has been and is translated again into a necessity where we find ourselves continuously redefining theatricality and shifting the boundaries between media practices.

Our inability to lastingly define genres or media – to be precise, we are obliged to remain true to a defined conjuncture (5th century (BC) Greek tragedy, the French theater of the *Grand Siècle*, American vaudeville theater, etc.) – is the most striking display of this. And technological progress derived from the mastery of electricity and, more recently, from digital developments, have only accentuated and rendered again more apparent this changing media reality that brings into light the very porous frontiers (between media practices) or which, more precisely, reveal their invented nature (frontiers are above all discursive constructions). The audiences that attend the most recent theater productions and dance shows in fact regularly experience this: we find many more shared characteristics between these two reputed and distinct art practices than between current theater and that of the 17th century. Ontological discourse traces frontiers, suggests historical filiations, deduces theoretical ties between practices of different periods, that the stage reality enjoys interfering with.[191]

If it is undeniable that the different elements which enter into what constitutes theatricality vary, it is equally true that these defining elements are not exclusive to theater practice. We must therefore understand theatricality as a conjuncturing ensemble wherein not only are none of its elements necessary and permanent, but not one of them is exclusive to theater. The frontality principle associated with the Italian stage, for example, is not unique to theater, as it also applies to cinema, and to the magic lantern. In the same way, the concept of presence, which Henri Gouhier has elevated to the 'essential' quality of theater, also applies to dance, as we have just mentioned, as well as to the circus and to concert performances, but also to sporting events (when one attends in person). And this same concept of presence – or of live – has paradoxically allowed radio and television to differentiate themselves from cinema and to win popular favor.

We unsurprisingly notice that even the term theatricality, as well as presence, only belatedly appeared. It began to be used at the beginning of the 1950s. This does obviously not mean that the very idea of the specificity of theater mediality did not exist prior, but that the perception we had of the situation meant that it was not useful to name it. Thomas Postlewait and Tracy C. Davis, who examined in detail the emergence of the concept in their book *Theatricality*,[192] thereby note that this concept was progressively built outside of theater. In intermedial terms, we would say that it is the remediation of certain elements associated with theater by extra-theatrical practices that lead to its opacification. In accordance with the principle of media *transparency* theorized Bolter and Grusin, theatricality, to the theater, was naturalized and passed unnoticed. It is not rendered opaque and it therefore becomes apparent only in migrating outside of the theater. Face to face with other forms of communicational and media practices, theatricality becomes salient, that is, very audible and very visible. Having thereby associated the theater with magnification, with exaggeration, with masking, with padding, with lying that is 'with the perverted, the artificial, the unnatural, the abnormal, the decadent, the effete, the diseased',[193] that which standard language understands by 'making theater'; or when there is the addition of the epithet 'theatrical' to the substantives gesture, pose, attitude, situation, or conclusion, etc.

---

191 This is what we described in chapter 4.
192 Tracy C Davis and Thomas Postlewait, *Theatricality*, Cambridge; New York: Cambridge University Press, 2003, p. 13.
193 *Performativity and Performance*, ed. by Andrew Parker, Eve Kosofsky Sedgwick, and English Institute, Essays from the English Institute, New York: Routledge, 1995, p. 5.

The 'theatrical' – and through this the very idea of theatricality – therefore rapidly acquires a negative connotation which, by contamination, would extend to the theater itself. It would be separated from, according to Postlewait and David, a profound 'anti-theatrical prejudice' that undermines the very legitimacy of theater as a representational art. At the end of the 19th century, the arrival of realism to the stage can therefore be understood as a desire to reaffirm this legitimacy by putting an end to consistent opacifying of the processes of theater mediation and by rendering them transparent. This confirms that theatricality is tied to its historical period. We otherwise notice that the arrival of stage realism corresponds, in time, to a marked acceleration of research and of experimentation in the field of sound and image reproduction, that would give birth to the large electric media of the 20th century. In the mimetic logic that realist theater defends, theatricality does not disappear, but its effect of enlargement and of exaggeration diminishes to the benefit of a 'reliving' (of reality). The distance between the original and the reproduced therein achieves its most reduced form possible, according to the users of the period.

Bolter and Grusin remind us, however, that transparency is neither the ultimate objective of mediation nor an inevitable and desired outcome, as opposed to what the reproduction technology industry has advocated for more than a century. Opacity, as the two authors demonstrate well, is omnipresent in history and if, as a whole, the need to experience an immediate relationship with what is represented – thanks to transparency – seems dominant, our fascination for the performance of mediation in work – which is a way to define opacity – never entirely disappears. It is in fact a reversal of this tendency that we witnessed during the decline of realism and the appearance of the avant-gardes. The principle of opacity has therefore gained legitimacy, the avant-gardes even made theatricality the substance and the concept of their shows. It thereby opened an age of 'theatricalized theater' that we will no longer cease to celebrate, first in restrained circles from Dada to Meyerhold, from Artaud to Brecht, then on a larger scale with postmodernity. For Postlewait and Davis, this change set in motion the progressive slide from 'representational' to 'presentational' [c556] and brought to the fore the performative dimension of mediation. We could add that this also marked the end of the domination of the traditional mimetic regime.

Integrating the lessons of Mallarmé and Hegel, who sought to make language – and poetic language – an autonomous universe, a world in itself, the heavy theatrical machine therefore provided a self-sufficient stage universe. This negation of the referential world has conferred to theatricality an unequal status in history: it took on a positive value and was simultaneously the end and the means of performance. It also became a privileged object of study by semioticians, whose influence overrode the entire field of human sciences. The prestige of semiotics rebounded on theatricality and it is in this context, or because of it, that theater studies emerged. Until then, theater had been considered a literary genre and its study was principally limited to that of dramatic text and their authors. In the same way that presence had been recovered to impose an essentialist conception of theater, theatricality would rely on major institutional reform: we therefore note the emancipation of an integrated practice to a discipline, the rise of theater to the status of a completely separate field of university study. This brought out strange divides, including the separation of the object of study and the literary: the literarity (the text) confined to the departments of literature, theatricality (what is left) brought to theater studies. It is in this sense that we must understand the famous declaration of Roland Barthes, one of the most influential French critiques of the 1950s and 1960s, who is also one of the first theorists to use the word theatricality (from 1954): 'What is theatricality? It is theatre without the text, it is a density of signs and sensations that manifests on stage [...]'.[194]

From an intermedial point of view, this definition is obviously absurd.[195] Not only does the text participate in theatricality, but theatricality is also in the text in the sense that it is the product of a premediation, as Grusin uses the term: without exception, the text is in fact conditioned, or pre-formated, by the practical conditions of the stage representation in effect at the time of its writing. It is true that this idea, evoked by Barthes, of the progressive construction of meaning, that affects the domain of creation as much as that of reception, includes a major performative dimension (the term performativity had not yet entered into use in the mid 50s, at least in the theater domain), it is above all the removal on which Barthes's definition is founded – 'theatre without the text' – that made a lasting impression on our minds. It is also what provoked a profound crisis from which theater has not entirely recovered from six decades later.

With the avant-gardes and postmodernity burgeoning, the progressive abandon of the rep-resentational, in favor of the presentational, paved the way for a new approach marked by performativity and collaboration between practices – from interdisciplinary to multidisciplinary.

---

194 Roland Barthes, 'Le théâtre de Baudelaire', in *Essais critiques*, Collection points Essais, 127, Impr,
     Paris: Éd. du Seuil, 1995, p. 41, our translation.
195 Maurice Merleau-Ponty, *Phenomenology of Perception*, London, New York: Routledge, 1962, citing
     Proust's *The Guermantes Way*, talks about the actress Berma playing the role of Phaedra. At the
     heart of the analysis that he offers is a critique of the idea of theater as representation. The theater
     experience consists in once again rendering present an event that is no longer there. The actor would
     therefore represent a character, in a structure strongly opposing the subject and the object: an object
     represented and a subject representing. Merleau-Ponty's idea is, by contrast, that theater in no way
     engenders a representative structure. The relationship between actor and character can be considered
     as an explanatory example of what is produced in the chiasmus between visible and invisible. We are
     not before a visible center – once upon a time there was Phaedra – that is reproduced, represented
     here and now thanks to the actor's action. It is more a case of the invisible, and this invisibility
     structures all the visible; Phaedra is the invisible. What is produced on stage is a sort of embodiment:
     the actor becomes the visibility of the invisible; it is therefore Phaedra on stage and not Berma. Here we
     are close to Deleuze's statement according to which roles supersede actors; but, certainly, Merleau-
     Ponty's notion is more complex. In the classical structure, the spectator would be caught in between
     two poles: he would ask himself if he sees Phaedra (the before) or Berma (the after), if Berma is none
     other than that which brings Phaedra into existence or if Phaedra is not but a deferred manifestation
     of Berma, and it is therefore not only according to the latter that we can conceive of the former. But in
     the experience of theater we do not at all feel this tension and this polarity. We do, however, perceive
     a chiasmus, the encroachment of Phaedra towards Berma. The spectator is not able to see Berma all
     while knowing that the actress is in the process of playing the role of Phaedra; the spectator perceives
     the single unit of role and of actor. It is not possible to see Phaedra on stage without Berma, but we
     can neither perceive Berma without Phaedra. Proust realizes this when he asserts that whatever
     he had studied in advance on Phaedra's role, which in principle would be the common ground of
     all the actresses that play Phaedra, to be capable of subtracting it, of not gathering it as the residue
     of Mrs. Berma's talent, he notices that this talent is only made with the role.

It would have been logical that, in this context, the concept of theatricality would simply be enriched by a performative dimension, but two factors have inhibited this development. On the one hand, semiotic thought has had the effect of turning the attention of researchers from performativity to an in-between model (upon which the sign and mimesis are founded) which, in some regards, marks a regression. On the other hand, the marked interest of semioticians for theatricality had the paradoxical effect of disqualifying it: the loss of influence of semiotics renders the concept dubious and less respectable. But the cause of the decline was not related to the questions of identity evoked earlier: the appropriation of mediality (theatricality) by a discipline (theater studies) can only, in fact, bring us back to earlier identity reflexes in other disciplines, reigniting the essentialist flame: theatricality became the mark of theater expansionism and this remediation of the theater has been perceived as a threat to ontology in other practices, that is, as a 'betrayal'.

The virulent attack of the American historian and art critic, Michael Fried, against the 'theatralisation' of the visual artists, offers the most memorable illustration of this. In 1967, judging that visual artists were too excessively retracing theater's methods and mediating strategies, he cautioned against the grave dangers that this attitude runs in relation to their own practice. Michael Fried therefore calls for a rejection of all theatricality in painting and in other fine arts: 'The success, even the survival of the arts has come increasingly to depend on their ability to defeat theatre'.[196]

Fried's initiative bears witness to two converging phenomena whose effects would heavily weigh on the field of theater studies, as well as on theater more generally. On the one hand, this call for 'survival' can only be insured by the 'defeat' of theater, that is by the repression of theatricality from within the field of theater. On the other hand, while the concept of theatricality seems to be diverted from its most negative connotations and benefitted, thanks to the prestige of semiotics and to stage practices themselves, from unequal recognition, Fried attacked it not only by relying on old prejudices, but in underlining the outdated nature of the concept. By inviting minimalist artists to trade presence for presentness, inspired by phenomenological thinking, Fried confirms the triumph of the presentational and, indirectly, of the performative.

The attraction of the performativity concept derives largely from the successes won by this new artistic practice, that is, performance and the critical thought that it produces. But its most sudden breakthrough within theater studies and the hegemony that this concept rapidly exerted – and still exerts – is initially explained by the decline in the concept of theatricality and by the void that its rejection created (aggravated by the general decline in semiotic influence). Richard Schechner and his collaborators for the renowned journal *Performance Studies* have not waited to successfully apply the analytical principles of performance to analyze theater shows, thereby publishing a major shift in the course of theater studies. This 'performative turn' has profoundly enriched the approach of theater performance and its components, centering the observation precisely on the flow of actions that are produced and the agencies that are

196 Michael Fried, *Art and Objecthood: Essays and Reviews*, Chicago: University of Chicago Press, 1998, p. 18.

deployed in all the production phases of and during the theatrical event. To the criteria of fidelity so closely tied to the representational, the tenants of performativity have opposed this to the authenticity of the experience of the performer, of course, but also that of the viewer-experiencer-user, rendered co-creator. He separates himself, on the one hand, from the image of the performer who (would) play his life; on the other hand, he distances himself from that of the actor who (would) represent one. Even if reality is not so simple – because it always remains a part of the representational in any stage show and in its components, because mimesis has never entirely disappeared – this double equation remains.

In this era of the presentational, which was also an era of a 'new sharing of the sensible' as Jacques Rancière[197] describes it, we have approached the theater performance not only as an autonomous object to which we were exogenous observers: we view theater as an event in which we participate, to the same extent as the 'observer' that Karen Barad[198] evokes, measuring the quantum 'events' for which it is an implicated member. A question therefore arises: if this new sharing of the sensible that Rancière evokes is necessarily performative, does there exist a specific performative mode for the theater? And if so, how do we characterize it?

This question obviously echoes that of the distinction between mediating conjunctures, or between the ensembles of mediating conjunctures, that we have discussed in chapter 3. Josette Féral, who has successively been one of the influential theorists of theatricality and performativity, very recently offered one possible response:

> To again interrogate the concept of theatricality [...] in light of the multiple developments of the performative that characterize a contemporary theatre stage in constant change is a must.[199]

This suggestion is more than appropriate. By opposing theatricality and performativity, and by privileging the latter at the expense of the former – under the pretext that the latter is true and the former is false or because the former is weighed down by a heavy negative connotations (from enlargement and falsity to the semiotic aura), or even because the latter continues to enjoy an immense prestige within the human sciences – theater studies have perhaps not but retraced the model of old binary categories that they themselves have endured. Are we thereby still a part of the presentational era? Things evolve. We observe a marked return of the character, that of the narrative and, with it, that of mimesis obviously in renewed but recognizable forms. If we think for example about specific processes of documentary theater or of the use and nature of visual and sound projections that henceforth accompany all productions. We can hardly imagine, today, a theater performance without a soundtrack and video projections have entered into stage use.

197 Jacques Rancière, *Le partage du sensible: esthétique et politique*, Paris: Fabrique: Diffusion Les Belles Lettres, 2000.

198 Among other texts by Karen Barad, see Barad, *Meeting the Universe Halfway: Quantum Physics and the Entanglement of Matter and Meaning.*

199 Josette Feral, 'Les paradoxes de la théâtralité', *Théâtre/Public*, 2012, p. 8 https://hal-univ-paris3. archives-ouvertes.fr/hal-01497261 [accessed 12 June 2019], our translation.

To the question, 'does a theatrical performativity exist?', we respond 'yes'. All mediation is performative and, of course, a theatrical performativity does exist, and is constantly redefined. This theatrical performativity stems from an ensemble of mediating conjunctures that we call theatricality. Performativity is therefore neither the successor nor the substitute of theatricality; it is a component that is central and necessary because there is no mediation that is not performative. What distinguishes this performativity from other performativities is its theatricality, it is the ensemble of mediating conjunctures to which it is attached and which, in the best of cases, results in a theater show.

Conclusion

The results of our analyses reinforce the hypothesis underpinning the notion of mediating conjunctures: it is impossible to identify a single and determined mediating conjuncture in abstracting other conjunctures which participate in the mediating process. Each act identifying the particularities of a mediation situation consists in 'isolating' mediating conjunctures, which are dynamic and have no frontier. The place of representational and performative paradigms, in the theater and in the digital turn, offers, in this regard, an illuminating an example.

In this book, we have highlighted the increased strength of the performative paradigm at the expense of the representational paradigm. From an essentialist point of view, we could believe that we are living in an era when mediation forms are therefore characterized by this paradigm and that there is no longer a place for mimesis and for representation. And yet, the reality is more complex: identifying the actions tied to each of these paradigms only occurs as a result of a particular kind of observation of mediating conjunctures. This observation consists in isolating a vast movement in order to focus our attention on only one part of their complex and open structure.

The 'digital turn' and the 'theater' therefore form 'parts' of mediating conjunctures which, in themselves, are not separated into discrete units. We can identify reoccurring forms in what makes up these conjunctures and, among its forms, it is in fact possible to notice a presence and a performative tendency. But we can just as well also turn our gaze to other elements and other reoccurrences. In this case, for example, we could associate the return of a character and of a narrative, in the theater domain, to the rapid development of virtual reality technology for fictional or entertainment purposes. We could therefore identify the emergence of a contradictory tendency to that of the performative paradigm, according to which mimesis is destined for a strong comeback due to the necessity of preserving the very possibility of producing a diegesis. So that one may tell a story, it is fundamental that the story's world be clearly separated from the world in which the reader or the spectator find themselves: it is necessary to put the reader and the spectator in a position to receive a representation.

The contribution of the notion of mediating conjunctures offers the possibility of identifying patterns and attributing them to an ontological meaning. This refers not only to interpretations of reality: it is reality itself that is open to observation. But at the same time, reality is plural because it is made from mediating conjunctures that are simultaneously the context and the outcome of the actions of elements that are involved.

The intermedial approach, in this sense, merges in an attempt to bring together agency and structure, constructivism and realism, which seems to be, in our time, the primary objective of philosophy.

BIBLIOGRAPHY

Aarseth, Espen J., *Cybertext: Perspectives on Ergodic Literature*, Baltimore, Md: Johns Hopkins University Press, 1997.

Acland, Charles R, *Residual Media*, Minneapolis: University of Minnesota Press, 2007.

Agamben, Giorgio, *Qu'est-ce qu'un dispositif?*, Paris: Payot & Rivages, 2007.

——, *What Is an Apparatus? And Other Essays*, Meridian, Crossing Aesthetics, Stanford, Calif: Stanford University Press, 2009.

Andreazza, Fabio, 'La conversion de Pirandello au cinéma', *Actes de la recherche en sciences sociales*, 161-162.1 (2006), 32 https://doi.org/10.3917/arss.161.0032.

Aristotle, *Physics*, ed. by C. D. C Reeve, 2018.

——, *The Metaphysics*, trans. by Hugh Tredennick, Cambridge, Mass.: Harvard University Press, 1989.

Arnheim, Rudolf, *Film Als Kunst.*, Frankfurt am Main: Suhrkamp, 2002.

Auslander, Philip, *Liveness: Performance in a Mediatized Culture*, London; New York: Routledge, 2011.

——, *Presence and Resistance: Postmodernism and Cultural Politics in Contemporary American Performance*, Annotated edition edition, Ann Arbor: University of Michigan Press, 1994.

Austin, John Langshaw, *How to Do Things with Words: [The William James Lectures Delivered at Harvard University in 1955]*, ed. by James Opie Urmson, 2. ed., [repr.], Cambridge, Mass: Harvard Univ. Press, 2009.

Bachimont, Bruno, 'Le numérique comme support de la connaissance: entre matérialisation et interprétation', in *Ressources vives. Le travail documentaire des professeurs en mathématiques*, ed. by Ghislaine Gueudet et Luc Trouche, Paideia, PUR et INRP, 2010, pp. 75-90 http://hal.archives-ouvertes.fr/hal-00496590 [accessed 12 November 2012].

——, *Le sens de la technique: Le numérique et le calcul*, 2010.

Barad, Karen, *Meeting the Universe Halfway: Quantum Physics and the Entanglement of Matter and Meaning*, Second Printing edition, Durham: Duke University Press Books, 2007.

——, 'Posthumanist Performativity: Toward an Understanding of How Matter Comes to Matter', *Signs: Journal of Women in Culture and Society*, 28.3 (2003), 801-31 https://doi.org/10.1086/345321.

Barbier, Frédéric, and Catherine Bertho-Lavenir, *Histoire des médias: de Diderot à Internet*, Paris: Colin, 1996.

Barthes, Roland, *Le bruissement de la langue*, Seuil, 1993.

——, 'Le théâtre de Baudelaire', in *Essais critiques*, Collection points Essais, 127, Impr, Paris: Éd. du Seuil, 1995.

Baudrillard, Jean, *Le crime parfait*, Paris: Editions Galilée, 1995.

Benjamin, Walter, *Illuminations: Essays and Reflections*, ed. by Hannah Arendt, trans. by Harry Zohn, New York: Schocken Books, 1969.

Bishop, Robert C., 'Determinism and Indeterminism', *arXiv:Physics/0506108*, 2005 http://arxiv.org/abs/physics/0506108 [accessed 28 May 2018].

Boenish, Peter, 'Mediation Unfinished: Choreographing Intermediality in Contemporary Dance Performance', in *Intermediality in Theatre and Performance*, ed. by Freda Chapple and Chiel Kattenbelt, Amsterdam; New York: Rodopi, 2006, pp. 151-66.

Bolter, J. David, and Richard A Grusin, *Remediation: Understanding New Media*, Cambridge, Mass.: MIT Press, 2000.

Braidotti, Rosi, *The Posthuman*, Cambridge, UK; Malden, MA, USA: Polity Press, 2013.

Butler, Judith, *Excitable Speech: A Politics of the Performative*, New York: Routledge, 1997.

——, *Gender Trouble: Feminism and the Subversion of Identity*, 1 edition, New York: Routledge, 2006.

Cambron, Micheline, and Marilou St-Pierre, 'Presse et ondes radiophoniques', *Sens Public*, 2016 http://www.sens-public.org/article1199.html [accessed 29 May 2018].

Chapple, Freda, Chiel Kattenbelt, International Federation for Theatre Research, and Theatre and Intermediality Working Group, *Intermediality in Theatre and Performance*, Amsterdam; New York: Rodopi, 2006.

Clüver, Claus, 'Intermediality and Interarts Studies', in *Changing Borders: Contemporary Positions in Intermediality*, ed. by Jens Arvidson, Lund: Intermedia Studies Press, 2007, pp. 19-37.

Cotton, Nicholas, 'Du performatif à la performance', *Sens Public*, 2016 http://www.sens-public.org/article1216.html [accessed 20 February 2018].

Cowley, Malcolm, ed., *Writers at Work: The Paris Review Interviews: First Series*, New York: Penguin Books, 1977.

Cusset, François, *French Theory: How Foucault, Derrida, Deleuze, & Co. Transformed the Intellectual Life of the United States*, Minneapolis, Minn. [u.a.: Univ. of Minnesota Press, 2008.

Davis, Tracy C, and Thomas Postlewait, *Theatricality*, Cambridge; New York: Cambridge University Press, 2003.

De Landa, Manuel, *Assemblage Theory*, Speculative Realism, Edinburgh: Edinburgh University Press, 2016.

Deleuze, Gilles, *Cinéma*, Paris: Éditions de Minuit, 1983.

Déotte, Jean-Louis, *Appareils et formes de la sensibilité*, L'Harmattan, 2005.

Didi-Huberman, Georges, *La ressemblance informe: ou le gai savoir visuel selon Georges Bataille*, 2019.

Diels, Hermann, Walther Kranz, and Kathleen Freeman, *The Presocratic Writings*, 2017 http://pm.nlx.com/xtf/view?docId=presocratics_gr/presocratics_gr.00.xml [accessed 18 February 2019].

Dolar, Mladen, *A Voice and Nothing More*, Cambridge, Mass: The MIT Press, 2006.

Doueihi, Milad, *Digital Cultures*, Cambridge, Mass.: Harvard University Press, 2011.

Dworkin, Craig, *No Medium*, 1st MIT Press paperback edition, Cambridge, Mass London: MIT Press, 2015.

Dyens, Ollivier, *Enfanter l'inhumain: le refus du vivant*, Montréal: Triptyque, 2012.

Eisentein, Elizabeth L, *The Printing Revolution in Early Modern Europe*, Cambridge: Cambridge University Press, 1983.

Elleström, Lars, *Media Borders, Multimodality and Intermediality*, Basingstoke [u.a.: Palgrave Macmillan, 2010.

——, *Media Transformation: The Transfer of Media Characteristics Among Media*, Houndsmill, Basingstoke, Hampshire New York: Palgrave Macmillan, 2014.

Faccini, Dominic, Georges Bataille, Michel Leiris, Carl Einstein, and Marcel Griaule, 'Critical Dictionary', *October*, 60 (1992), 25-31 https://doi.org/10.2307/779027.

Feral, Josette, 'Les paradoxes de la théâtralité', *Théâtre/Public*, 2012 https://hal-univ-paris3.archives-ouvertes.fr/hal-01497261. [accessed 12 June 2019]

Ferraris, Maurizio, *Documentality: Why It Is Necessary to Leave Traces*, trans. by Richard Davies, 1 edition, New York: Fordham University Press, 2012.

Foucault, Michel, 'Qu'est-ce qu'un auteur?', in *Dits et écrits*, Paris: Gallimard, 1994.

Foucault, Michel, Daniel Defert, François Ewald, and Jacques Lagrange, *Dits et écrits, 1954-1988. II, II*, [Paris]: Gallimard, 2001.

Fried, Michael, *Art and Objecthood: Essays and Reviews*, Chicago: University of Chicago Press, 1998.

Galloway, Alexander R, *The Interface Effect*, Cambridge, UK; Malden, MA: Polity, 2012.

Galloway, Alexander R., Eugene Thacker, and McKenzie Wark, eds., *Excommunication: Three Inquiries in Media and Mediation*, Trios, Chicago; London: The University of Chicago Press, 2014.

Gaudreault, André, and Philippe Marion, *The End of Cinema? A Medium in Crisis in the Digital Age*, New York: Columbia University Press, 2015.

Gitelman, Lisa, *Always Already New: Media, History, and the Data of Culture*, The MIT Press, 2008.

Grusin, R., *Premediation: Affect and Mediality After 9/11*, 2010 edition, Basingstoke England; New York: Palgrave Macmillan, 2010.

Grusin, Richard, 'Radical Mediation', *Critical Inquiry*, 42.1 (2015), 124-48 https://doi.org/10.1086/682998.

Herzogenrath, Bernd, ed., *Travels in Intermedia[lity]: Reblurring the Boundaries*, Interfaces: Studies in Visual Culture, 1st [ed.], Hanover: Dartmouth College Press, 2012.

Huhtamo, Erkki, and Jussi Parikka, eds., *What Is Media Archeology*, Cambridge: Polity Press, 2015.

Jenkins, Henry, *Convergence Culture: Where Old and New Media Collide*, Revised edition, New York: NYU Press, 2008.

Junod, Philippe, *Transparence et opacité: Essai sur les fondements théoriques de l'art moderne. Pour une nouvelle lecture de Konrad Fiedler*, Nîmes: Chambon, 2004.

Kattenbelt, Chiel, 'Theatre as the Art of the Performer and the Stage of Intermediality', in *Intermediality in Theatre and Performance*, ed. by Freda Chapple and Chiel Kattenbelt, Amsterdam; New York: Rodopi, 2006.

Kittler, Friedrich, *Optical Media*, trans. by Anthony Enns, 1 edition, Cambridge: Polity, 2009.

Lawrence, Fred, *The Beginning and the Beyond: Papers from the Gadamer and Voegelin Conferences*, Chico: Calif.: Scholars Press, 1984.

Lovink, Geert, *Dark Fiber: Tracking Critical Internet Culture*, Electronic Culture-History, Theory, Practice, Cambridge, Mass: MIT Press, 2002.

Marion, Philippe, *L'année des médias 1996*, 1997 https://dial.uclouvain.be/pr/boreal/object/boreal:83077 [accessed 16 February 2019].

Marvin, Carolyn, *When Old Technologies Were New: Thinking About Electric Communication in the Late Nineteenth Century*, Nachdr., III, 1990.

McAuley, Gay, 'The Video Documentation of Theatrical Performance', *New Theatre Quarterly*, 10.38 (1994), 183-94 https://doi.org/10.1017/S0266464X00000348.

MCLuhan, Marshall, *Understanding Media*, New York: Signet, 1966.

Merleau-Ponty, Maurice, *Phenomenology of Perception*, London, New York: Routledge, 1962.

Méchoulan, Eric, 'Intermédialité, ou comment penser les transmissions', *Fabula Colloques*, 2017 http://www.fabula.org/colloques/document4278.php [accessed 29 May 2018].

Méchoulan, Éric, 'Intermédialités: le temps des illusions perdues', *Intermédialités : Histoire et théorie des arts, des lettres et des techniques / Intermediality : History and Theory of the Arts, Literature and Technologies*, 1, 2003, 9-27 https://doi.org/https://doi.org/10.7202/1005442ar.

Monjour, Servanne, *Mythologies postphotographiques: l'invention littéraire de l'image numérique*, Montréal: PUM, 2018 http://parcoursnumeriques-pum.ca/introduction-158.

Moser, Water, 'L'interartialité: pour une archéologie de l'intermédialité', in *Intermedialité et socialité: histoire et géographie d'un concept*, ed. by Marion Froger and Jürgen E and Müller, Münster: Nodus, 2007, pp. 69-92.

Murray, Janet Horowitz, *Hamlet on the Holodeck: The Future of Narrative in Cyberspace*, Cambridge, Mass: MIT Press, 1998.

Müller, Jürgen, *Intermedialität: Formen Moderner Kultureller Kommunikation*, Münster: Nodus Publikationen, 1996.

Nelson, T. H., 'Complex Information Processing: A File Structure for the Complex, the Changing and the Indeterminate', in *Proceedings of the 1965 20th National Conference*, ACM '65, New York, NY, USA: ACM, 1965, pp. 84-100 https://doi.org/10.1145/800197.806036.

Norman, Donald A., *The Design of Everyday Things*, 1st Basic paperback, New York: Basic Books, 2002.

Ortel, Philippe, 'Note sur une esthétique de la vue: Photographie et littérature', 2002 https:// doi.org/http://isidore.science/document/10.3406/roman.2002.1164.

Ortiz, Darwin, *The Annotated Erdnase*, Pasadena, Calif: Magical Publications, 1991.

Parker, Andrew, Eve Kosofsky Sedgwick, and English Institute, eds., *Performativity and Performance*, Essays from the English Institute, New York: Routledge, 1995.

Pelisson, *Regie Theatrale et Mise En Scene*, Villeneuve d'Ascq, France: Presses Universitaires du Septentrion, 2014.

Phelan, Peggy, *Unmarked: The Politics of Performance*, 1 édition, London; New York: Routledge, 1993.

Pierce, John R., *An Introduction to Information Theory: Symbols, Signals & Noise*, 2nd, rev. ed, New York: Dover Publications, 1980.

Plato, *Parmenides*, trans. by Benjamin Jowett, Adelaide: The University of Adelaide Library, 2008 https://en.wikisource.org/wiki/Parmenides [accessed 21 December 2018].

Plimmer, Gill and Daniel Thomas, 'Network Rail's Fibre Optic Network Attracts Telecoms Interest', 2016 https://www.ft.com/content/173b6c06-f1da-11e5-aff5-19b4e253664a [accessed 10 April 2018].

Popper, Karl R., and Giancarlo Bosetti, *The Lesson of This Century: With Two Talks on Freedom and the Democratic State*, London; New York: Routledge, 2000.

Pougin, Arthur, *Dictionnaire Historique Et Pittoresque Du Théâtre Et Des Arts Qui s'y Rattachent: Poétique, Musique, Danse, Pantomime, Décor, Costume, Machinerie, ... Fètes Publiques, Réjouissa*, Forgotten Books, 2018.

Rancière, Jacques, 'Ce que "medium" peut vouloir dire: l'exemple de la photographie', *Appareil*, 1, 2008 https://doi.org/10.4000/appareil.135.

———, *Le partage du sensible: esthétique et politique*, Paris: Fabrique: Diffusion Les Belles Lettres, 2000.

———, *Le spectateur émancipé*, La Fabrique éditions, 2008.

Reason, M, *Documentation, Disappearance and the Representation of Live Performance*, Place of publication not identified: Palgrave Macmillan, 2016.

Ricoeur, Paul, *Memory, History, Forgetting*, trans. by Kathleen Blamey and David Pellauer, New edition édition, Chicago, Ill.: University of Chicago Press, 2006.

Ricoeur, Paul, *Soi-même comme un autre*, POINTS, 2015.

Rose, Mark, *Authors and Owners: The Invention of Copyright*, Cambridge, Mass.: Harvard University Press, 1993.

Rykner, Arnaud, *Le tableau vivant et la scène du corps: vision, pulsion, dispositif*, Hyper Article en Ligne – Sciences de l'Homme et de la Société, 2013 https://isidore.science/document/10670/1.0jcfgt [accessed 16 February 2019].

Schaeffer, Pierre, *Traité Des Objets Musicaux: Essai Interdisciplines*, Paris: Seuil, 2002.

Sibony, Daniel, *Entre-deux: l'origine en partage*, Paris: Editions du Seuil, 2003.

Smith, Jacob, and American Council of Learned Societies, *Vocal Tracks: Performance and Sound Media*, Berkeley: University of California Press, 2008 https://login.gbcprx01.georgebrown.ca/login?url=https://ebookcentral.proquest.com/lib/georgebrown-ebooks/detail.action?docID=358943 [accessed 16 February 2019].

Soleri, Paolo, *Arcologie: la ville à l'image de l'homme*, Roquevaire: Parenthèses, 1980.

Sterne, Jonathan, *The Audible Past: Cultural Origins of Sound Reproduction*, Durham [N.C.: Duke university press, 2003.

Svensson, Patrik, 'Envisioning the Digital Humanities', 6.1 (2012) http://www.digitalhumanities.org/dhq/vol/6/1/000112/000112.html [accessed 5 December 2014].

Tanner, Jane, 'New Life for Old Railroads; What Better Place to Lay Miles of Fiber Optic Cable', *The New York Times*, 2000 https://www.nytimes.com/2000/05/06/business/new-life-for-old-railroads-what-better-place-to-lay-miles-of-fiber-optic-cable.html [accessed 10 April 2018].

Terranova, Tiziana, *Network Culture: Politics for the Information Age*, London; Ann Arbor, MI: Pluto Press, 2004.

Thacker, Eugene, 'Dark Media', in *Excommunication: Three Inquiries in Media and Mediation*, ed. by Alexander R. Galloway, McKenzie Wark, and Eugene Thacker, Trios, Chicago; London: The University of Chicago Press, 2014.

Treleani, Matteo, 'Le spectre et l'automate. Deux figures du spectateur', in *D'un écran à l'autre. Les mutations du spectateur*, Delavaud, G. Chateauvert, J., Harmattan, 2016.

Véron, Eliséo, 'De l'image sémiologique aux discursitivtés. Le temps d'une photo', *Hermès*, 13, 1994, 45 https://doi.org/10.4267/2042/15515.

Véron, Éliséo, 'De l'image sémiologique aux discursivités, From the semiotic image to discursivity', *Hermès, La Revue*, 13, 2013, 45-64 http://www.cairn.info/resume.php?ID_ARTICLE=HERM_013_0045 [accessed 30 April 2017].

Vitali Rosati, Marcello, *S'orienter dans le virtuel*, Cultures numériques, ISSN 2118-1926, Paris, France: Hermann, 2012.

Wolf, Mark J. P, *Building Imaginary Worlds: The Theory and History of Subcreation*, New York: Routledge, New York, 2012.

www.ingramcontent.com/pod-product-compliance
Lightning Source LLC
Chambersburg PA
CBHW060623210326
41520CB00010B/1445